Nano-Interconnect Materials and Models for Next Generation Integrated Circuit Design

Aggressive scaling of device and interconnect dimensions has resulted in many low-dimensional issues in the nanometer regime. This book deals with various new-generation interconnect materials and interconnect modeling, and highlights the significance of novel nano-interconnect materials for 3D integrated circuit design. It provides information about advanced nanomaterials like carbon nanotube (CNT) and graphene nanoribbon (GNR) for the realization of interconnects, interconnect models, and crosstalk noise analysis.

Features:

- Focuses on materials and nanomaterials utilization in next-generation inter-connects based on carbon nanotubes (CNT) and graphene nanoribbons (GNR).
- Helps readers realize interconnects, interconnect models, and crosstalk noise analysis.
- Describes hybrid CNT- and GNR-based interconnects.
- Presents the details of power supply voltage drop analysis in CNT and GNR interconnects.
- Overviews pertinent RF performance and stability analysis.

This book is aimed at graduate students and researchers in electrical and materials engineering, and nano-/microelectronics.

Nano-Interconnect Materials and Models for Next Generation Integrated Circuit Design

Edited by
Sandip Bhattacharya, J. Ajayan and
Fernando Avila Herrera

CRC Press
Taylor & Francis Group
Boca Raton London New York

CRC Press is an imprint of the
Taylor & Francis Group, an **informa** business

MATLAB® is a trademark of The MathWorks, Inc. and is used with permission. The MathWorks does not warrant the accuracy of the text or exercises in this book. This book's use or discussion of MATLAB® software or related products does not constitute endorsement or sponsorship by The MathWorks of a particular pedagogical approach or particular use of the MATLAB® software

First edition published 2024
by CRC Press
6000 Broken Sound Parkway NW, Suite 300, Boca Raton, FL 33487-2742

and by CRC Press
4 Park Square, Milton Park, Abingdon, Oxon, OX14 4RN

CRC Press is an imprint of Taylor & Francis Group, LLC

ISBN: 9781032363813 (hbk)
ISBN: 9781032363820 (pbk)
ISBN: 9781003331650 (ebk)

DOI: 10.1201/9781003331650

Typeset in Times
by codeMantra

Contents

About the Editors

Sandip Bhattacharya received his Ph.D. (Eng.) degree from the Indian Institute of Engineering Science and Technology (IIEST), India, in 2017. From October 2017 to December 2020, he worked as a postdoctoral researcher at the HiSIM research center, Hiroshima University, Japan. He is currently working as an Associate Professor and Head of the Department of Electronics and Communication Engineering at SR University, Warangal, Telangana, India. His current research interests are nano device and interconnect modeling.

J. Ajayan received his B.Tech. degree in Electronics and Communication Engineering from Kerala University in 2009, and M.Tech. and Ph.D. degrees in Electronics and Communication Engineering from Karunya University, Coimbatore, India, in 2012 and 2017, respectively. He is an Associate Professor in the Department of Electronics and Communication Engineering at SR University, Telangana, India. He has published more than 100 research articles in various journals and international conferences. He has published two books, more than ten book chapters, and has two patents. He is a reviewer of more than 30 journals for various publishers. He was the Guest Editor for several of the special issues. His areas of interest are microelectronics, semiconductor devices, nanotechnology, RF integrated circuits, and photovoltaics.

Fernando Avila Herrera has worked in the field of academic and semiconductor industry. He has involved with the modeling and characterization of semiconductor devices, especially MOSFETs. Further, he has experience in device reliability modeling, model parameter extraction, Verilog-A, TCAD, and EDA tools. He also has experience in HiSIM family models for parameter extraction and physics modeling and FPGA programming. He has collaborated with different groups for developing compact models.

Contributors

Syed Musthak Ahmed
Department of Electronics &
  Communication Engineering
SR University
Warangal, India

P. Anuradha
Department of Electronics &
  Communication Engineering
Chaitanya Bharathi Institute of
  Technology
Hyderabad, India

Mounika Bandi
Department of Electronics and
  Communication Engineering
SR University
Warangal, India

Sandip Bhattacharya
Department of Electronics and
  Communication Engineering
SR University
Warangal, India

Shivangi Chandrakar
Department of Electronics and
  Communication Engineering
International Institute of Information
  Technology, Naya Raipur
Naya Raipur, India

Debaprasad Das
Department of Electronics and
  Communication Engineering
Assam University
Silchar, India

Subhajit Das
HiSIM Research Center
SR University
Warangal, India

L. M. I. Leo Joseph
Department of Electronics &
  Communication Engineering
SR University
Warangal, India

Sayan Kanungo
Department of Electrical & Electronics
  Engineering
Birla Institute of Technology and
  Science-Pilani, Hyderabad
Hyderabad, India

Bhawana Kumari
Department of Electronics Engineering,
  IIT(ISM), Dhanbad
Dhanbad, India

Manoj Kumar Majumder
Department of Electronics and
  Communication Engineering
International Institute of Information
  Technology, Naya Raipur
Naya Raipur, India

Sheshikala Martha
Department of Electronics &
  Communication Engineering
SR University
Warangal, India

Sangeeta Jana Mukhopadhyay
Department of Electrical & Electronics
  Engineering
Dr. Sudhir Chandra Sur Institute of
  Technology and Sports Complex
Kolkata, India

Shantikumar Nair
Amrita School of Nanosciences and
  Molecular Medicine
Amrita Vishwa Vidyapeetham
Kochi, Kerala, India

Hafizur Rahaman
School of VLSI Technology
Indian Institute of Engineering Science
  and Technology, Howrah
Howrah, India

Ch. Rajendra Prasad
Department of Electronics &
  Communication Engineering
SR University
Warangal, India

Shashank Rebelli
Department of ECE, SR University
Warangal, India

Manodipan Sahoo
Department of Electronics Engineering,
  IIT(ISM), Dhanbad
Dhanbad, India

Laxman Raju Thoutam
Amrita School of Nanosciences and
  Molecular Medicine
Amrita Vishwa Vidyapeetham
Kochi, Kerala, India

Aditya Tiwari
Department of Electrical & Electronics
  Engineering
Birla Institute of Technology and
  Science-Pilani, Hyderabad
Hyderabad, India

Santasri Giri Tunga
School of VLSI Technology
Indian Institute of Engineering Science
  and Technology, Howrah
Howrah, India

Wen-Sheng Zhao
School of Electronics and Information
Hangzhou Dianzi University
Hangzhou, China

1 Nanomaterials for Next-Generation Interconnects

Aditya Tiwari, Sangeeta Jana Mukhopadhyay, and Sayan Kanungo

1.1 INTRODUCTION

1.1.1 BACKGROUND

Owing to their augmented electrical, optical, mechanical, chemical, and thermal properties, nanomaterials have proven to be superior to their bulk counterpart in almost every aspect of consumer applications. For the past two decades, a great deal of scientific study is going on exploring these materials in order to take advantage of these nanomaterials in various fields of research. By changing the type of orbital hybridization, carbon can take the form of various nano-allotropes. Since all nano-allotropes of carbon have unusual physical and chemical characteristics, they are widely used in many different applications, particularly in the electronics industry. Fullerenes, graphene, and carbon nanotubes/nanoribbons are some of the most common examples of three-, two-, and one-dimensional carbon nano-allotropes. These carbon-based materials that have displayed such inherent properties can be easily exploited in the making of cutting-edge technology applications. Some of the unique characteristics of commonly used carbon-based nanomaterials are highlighted in Table 1.1 [1,2].

The aforementioned distinctive qualities suggest that carbon nanotubes (CNTs) and graphene nanoribbons (GNRs) may be used as replacement materials for future

TABLE 1.1

Comparison of Various Properties of Carbon Nano-Allotropes Significant in Interconnect Application

Property	Copper	Tungsten	Carbon Nanotube	Graphene/ Nanoribbon
Thermal conductivity (W/m K)	385	173	1,750–5,800	3,000–5,000
Maximum current density(A/cm^2)	10^7	10^8	$>10^9$	$>10^9$
Melting point (K)	1,357	3,695	3,800	3,800
Tensile strength (GPa)	0.22	1.51	11–63	130
Temperature coefficient of resistance (×10^{-3}/K)	4	4.5	−1.37 to 1.1	−1.47

DOI: 10.1201/9781003331650-1

nano-scale interconnects, thus improving electrical performance and removing electro-migration reliability issues that have been afflicting Cu-based nano-scale interconnects for a long time. In this chapter, the distinctive characteristics of various nanomaterials are highlighted, along with their current and potential interconnect applications in the future.

1.1.2 INTRODUCTION TO CARBON NANOMATERIALS

With a valency of four, carbon atoms can make single, double, and triple covalent bonds with other elements as well as with other carbon atoms. Additionally, they are capable of polymerizing, which is the process of creating long chains of atoms. Despite having identical chemical content, carbon atoms can exhibit different physical shapes and different physical attributes due to their unique electronic structure and atomic size as shown in Figure 1.1. With a small band gap between their $2s$ and $2p$ electronic shells, carbon atoms are capable of sp, sp^2, and sp^3 hybridizations. For example, graphite possesses sp^2 hybridization, while diamond has sp^3 hybridization [3–5]; see Figure 1.2. The primary criterion for classifying nanomaterials is the geometrical structure of the particles. The particles may be in the form of strips, spheres, or tubes. Carbon nanotubes (CNTs) are tube-shaped allotropes; fullerenes contain spherical or ellipsoidal nanoparticles, while graphene nanoribbons are made of strips of carbon atoms that are no longer than 100 nm in length [6–8].

Additionally, the highly delocalized electronic structure of sp^2-hybridized carbon nanomaterials points to the potential of these materials as high-mobility electronic materials. Customizing optical and optoelectronic qualities is made possible by the ability to adjust the band gap of semiconducting CNTs by controlling their diameter. Due to these factors, silicon and other traditional semiconducting materials are frequently replaced with carbon nanomaterials (CNMs) in electrical and optoelectronic applications [9].

Due to their low toxicity and widespread production for use, these carbon nanomaterials (CNMs), particularly carbon nanotubes (CNTs), graphene, graphene nanoribbons (GNRs), and carbon dots (CDs), find extensive technical applications in micro- and nano-electronics, gas storage, production of conductive plastics, composites, paints, textiles, batteries with extended lifetimes, biosensors, etc. [10–12].

1.1.3 CARBON NANOMATERIALS

1.1.3.1 Graphene

Graphene is a two-dimensional (2D) analog of graphite (or carbon) that has unique characteristics resulting from the bonding characteristics of carbon (C), which results in a sheet like planar structure, see Figure 1.3 [13]. This allotrope of carbon was first developed by A. Geim and Novoselov in 2004, which later won Novoselov a Nobel Prize in 2010 [14–16]. Three of the four valence electrons in C occupying the s, p_x, and p_y orbitals are involved in σ-bonding with three of its closest neighbors, forming

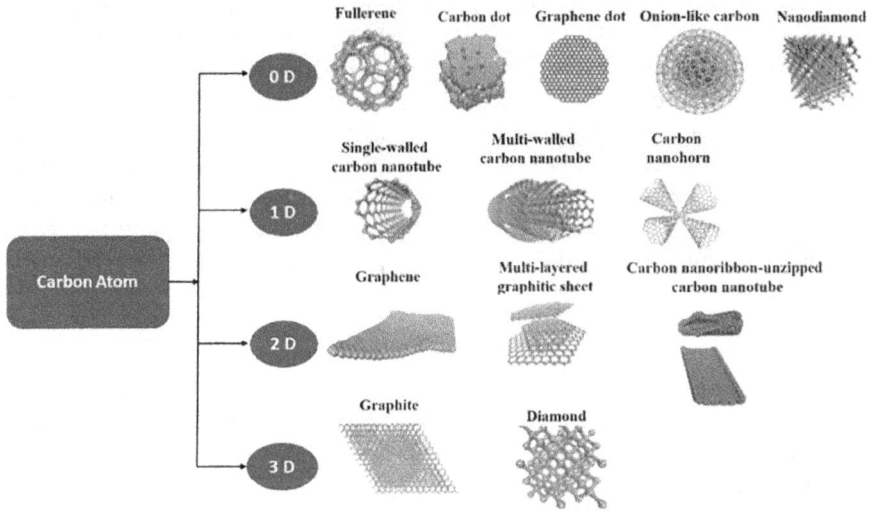

FIGURE 1.1   Classification of carbon nano-allotropes. 10.1021/cr500304f [59]

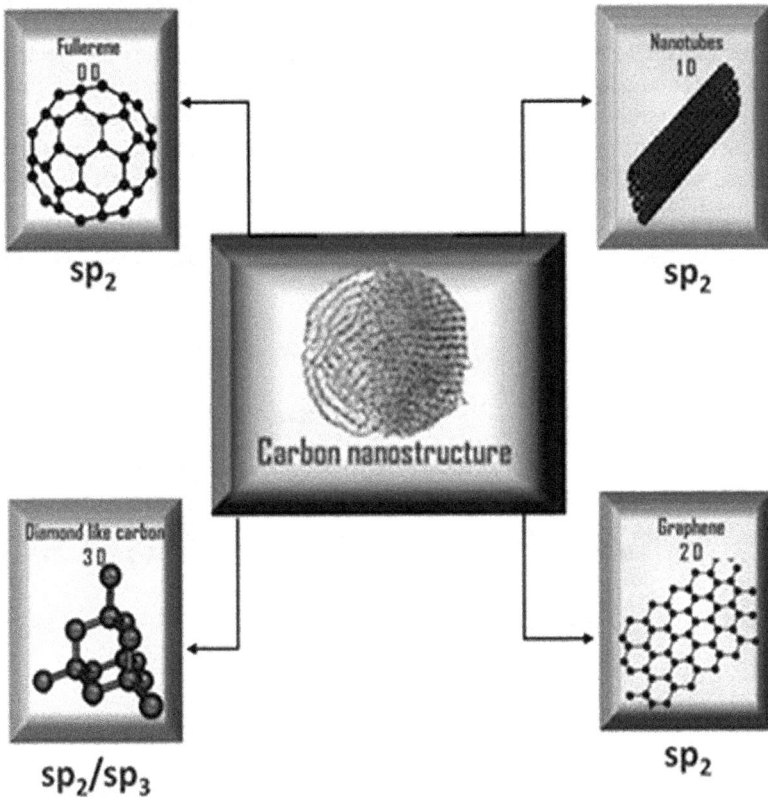

FIGURE 1.2   Hybridization of carbon nanostructure. 10.1007/978-3-030-02369-0_17 [60]

a honeycomb lattice structure. The remaining orbital (p_z), which is perpendicular to the one-dimensional (1D) sheet, is occupied by the fourth of these valence electrons, resulting in delocalized π-bonding and is responsible for the formation of a two-dimensional electron gas (2D-EG) with high mobility inside the sheets [17,18]. The delocalization of the π-bonding electrons allows for the graphene sheets to have very high mobility, up to 15,000–200,000 cm²/V s, which is only present in pristine graphene. Pristine graphene, however, has many drawbacks along with limitations in sensing applications; thus, various doping and modifications are employed in pristine graphene [19]. Furthermore, the synthesized graphene monolayer, however, is often contaminated with foreign particles or the substrate itself during synthesis process, which can limit its extraordinary electrical characteristics [20–24]. Because of this, generating and exploiting graphene for applications requiring high mobility and rapid processing raises serious concerns about cleanliness, grain size, and substrate interference. We are concentrating on the significant electrical properties of graphene in this review, but we should also highlight some of its other qualities to be thorough. A single flake of graphene will have a high breaking strength of about 40 N/m due to the 2D character of the material and the lack of slip planes linking the material's fracture strength to the solid's strong C–C bonding in a hexagonal ring [25]. The separation of electrons from phonons is another factor in the high thermal conductivity of 5,000 W/mK at ambient temperature. In addition to having a high breaking strength, graphene has a high Young's modulus of 1.0 TPa and a 20% elastic strain [25]. Potential uses for graphene include lightweight, thin, and flexible electric/photonics circuits and solar cells, as well as a number of improved or enabled industrial, medical, and chemical processes [26]. In addition, graphene can also be used for anti-corrosion coatings and paints, quicker and more efficient electronics, efficient and precise sensors, efficient solar panels, flexible displays, faster DNA sequencing, medicine delivery, and many more.

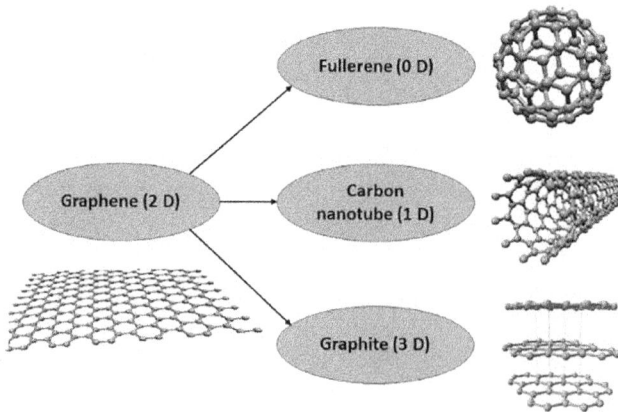

FIGURE 1.3   2D graphene and its other forms. [13]

1.1.3.2 Graphene Nanoribbons (GNRs)

Graphene nanoribbons (GNRs) are graphene strips with a width of less than 100 nm. They are also known as nano-graphene ribbons or nano-graphite ribbons. Initially, M. Fujita et al. proposed graphene-based ribbons as a theoretical model to investigate the edge and nano-scale size effect in graphene [27,28].

Graphene nanoribbons (GNRs) can be broadly categorized into three groups based on their edge structures: zigzag (ZZ), armchair (AC), and chiral GNRs; see Figure 1.4 [29]. Chiral ordered orientations are other ordered orientations that combine ZZ and AC edges. It is worth mentioning that the edge structures of GNRs have a significant impact on their attributes. All ZZ-edged GNRs (Z-GNRs), according to the nearest-neighbor tight binding hypothesis, are metallic [28–30]. Moreover, in Z-GNRs, conducting channels form close to the Fermi level as a result of the edge states concentrated at the ZZ edges that exhibit a ferromagnetic ordering [31,32]. Anti-ferromagnetism and bandgap opening occur in Z-GNRs when their width is narrow enough. Furthermore, the application of the electric field in the planar direction of Z-GNRs renders the nanoribbon half-metallic in nature, along with minimizing its spin degeneracy, utilizing them in spintronics applications [33–36]. On the other hand, the AC-edged GNRs (A-GNRs) exhibit alternating metallic and semiconducting behavior [31,32,37]. GNR has a wide variety of possible uses. The inclusion of GNRs in polymer hosts for the creation of innovative composite materials is the most prominent. GNRs possess the same high aspect ratio as multi-walled carbon nanotubes (MWCNTs), but changes in their nano-structure create surprising consequences. For instance, incorporating GNRs into a dielectric polymer host considerably modifies its electric properties [38,39] in a way that differs greatly from incorporating MWCNTs. The most surprising finding is that GNR-containing polymer composites have a remarkably low loss (0.02) at relatively high permittivity levels. This is significant because electronic component downsizing necessitates materials with high permittivity and low loss in the radio

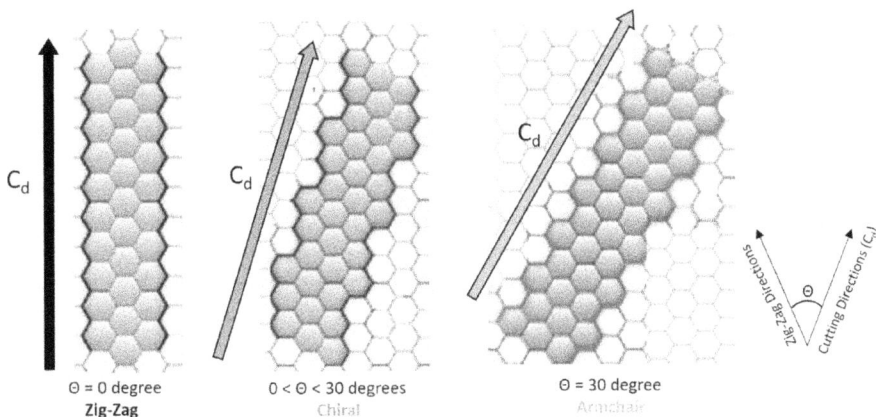

FIGURE 1.4 Classification of graphene nanoribbons. [29]

and low-frequency microwave ranges. Low loss is crucial in the high-frequency microwave range for antennas and other military applications. The loss and permittivity of composites may be tailored to suitable levels over a wide range by altering the type and amount of GNRs. The dielectric constant may be adjusted from moderate to extremely high (>1,000) values, while the loss tangent can be adjusted from ultralow to high, which are important in the development of various optical as well as electrical interconnects [39].

1.1.3.3   Carbon Nanotubes (CNTs)

Radushkevich and Lukyanovich together discovered tubular carbon nanostructures for the first time in 1952 [40]. Later on, Oberlin et al. reported the discovery of hollow carbon fibers with nanometer-sized dimensions [41]. Almost two decades later, Iijima published in the *Nature* journal crystal-clear photographs of carbon nanotubes (CNTs), sparking widespread attention [42]. Carbon nanotubes (CNTs) are one of the most popular carbon nanostructures, in which single-/multi-layer carbon (graphitic) sheets are rolled to form a cylindrical structure having a radius ranging from 0.7 to 50 nm [43] along with having lengths in micrometers [44]. In CNTs, each C-atom forms a bond with three other C-atoms with the help of covalent bond forming a hexagonal structure.

Recent technological developments have allowed the length of nanotubes to be measured in centimeters as well [45]. Based on how CNT sheets are wrapped into their tubes, CNTs have been divided into three distinct geometries named "zigzag" $(n, 0)$, "armchair" (n, n), and chiral (n, m); see Figure 1.5a and b [46–48]. The numbers (n, m) indicate how many steps there are along the carbon bonds in the hexagonal structure. If $n \neq m$, CNTs are regarded as chiral [46–49]. The spiral alignment (right- or left-handed) of the hexagonal rings through the axis of CNTs is indeed the source of the chirality [50]. Depending on their diameter and chirality, CNTs can exhibit either a semiconductor or metallic characteristic. The chiral vector, a pair of indices (n and m) that determines how the graphene sheet is rolled over, allows the CNTs to be armchair, zigzag, or chiral. Thus, the zigzag and chiral CNTs have electrical properties similar to semiconductors, including the SWCNTs (see Figure 1.5c) [47], and the armchair CNTs have electrical properties comparable to metals, including the MWCNTs (see Figure 1.5d) [47,51].

Carbon nanotubes have piqued the interest of researchers due to their simplicity and ease of manufacture. Many engineering applications have investigated the new features of nanostructured carbon nanotubes, such as large surface area, superior stiffness, and robustness. Due to their capabilities for aligned CNT growth, large-area deposition, and selective growth, chemical vapor deposition (CVD) technologies are best suited for developing CNTs for interconnect applications. MWCNTs, which can ensure metallic behavior, have been the main focus of the majority of CNT interconnect manufacturing research thus far. Carbon nanotube research has demonstrated applications in energy storage, hydrogen storage, electrochemical supercapacitors, field-emitting devices, transistors, nanoprobes and sensors, composite materials, templates including interconnects, and so on [52].

CNTs are typically categorized into three types as follows.

(a)

Chiral (m≠n)

Zigzag (n,0) Armchair (n,n)

(b)

Outer layer

Inner layer

(c)

Outer layer (layer 1)

Middle layer (layer 2) Inner layer (layer 3)

(d)

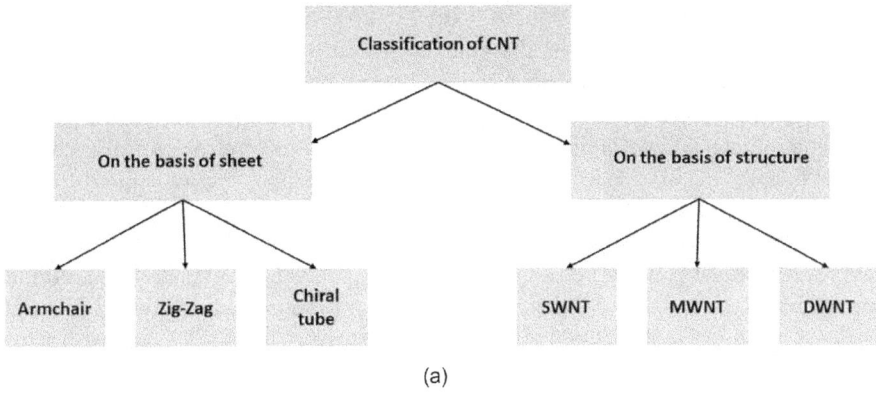

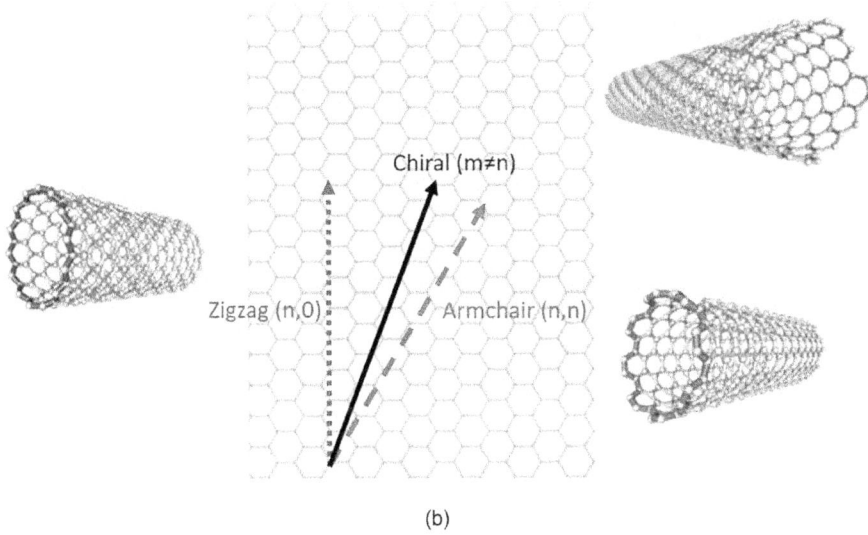

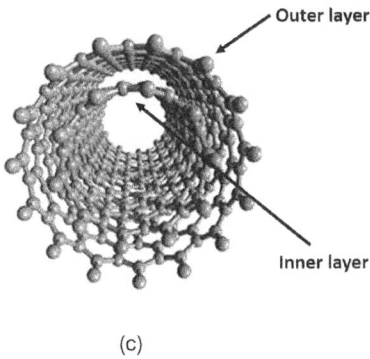

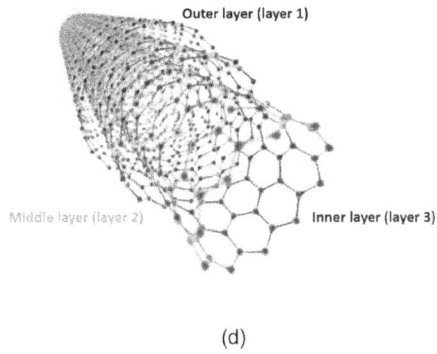

FIGURE 1.5 (a) Classification of carbon nanotubes (CNTs), (b) single-walled carbon nano-tubes (SWNTs) structure as a function of their chirality (zigzag, armchair, and chiral), (c) double-walled carbon nanotubes (DWNTs) structure, and (d) multi-walled carbon nanotubes (MWNTs) structure made with numerous concentric shells. [46–48, 51]

1.1.3.4 Single-Walled Carbon Nanotubes (SWCNTs)

A single graphene sheet with a diameter of 1–2 nm is rolled on to create single-walled CNTs [53]. SWCNTs have a tube length that can be millions of times greater than their diameter, which is close to 1 nm. The most significant examples of 1D nanomaterials with a dimension of 1–3 nm and dependent chirality characteristics are SWCNTs. As a semiconductor, SWCNTs have demonstrated to have exceptional thermal and mechanical characteristics, high mobility, high current-carrying efficiency, and negligible intrinsic capacitance. Additionally, SWCNTs have been extensively used as channel materials for radiofrequency (RF), digital, and macro-electronics [54]. By encasing a layer of graphite called graphene, which is one atom thick, in a special cylinder, the structure of a SWCNT can be systematically organized.

1.1.3.5 Double-/Triple-Walled Carbon Nanotubes (DWCNTs/TWCNTs)

The geometric link between SWCNTs and MWCNTs is produced by coaxially nesting two and three rolls of double- and triple-walled carbon nanotubes. In the samples, DWCNTs with inner diameters of between 1.0 and 2.0 nm predominate [55].

Two coaxial carbon nanotubes, with the outer tube encircling the inner tube, are used to create DWCNTs. The investigation of DWCNTs reveals the impact of inter-planar contact on the physical and chemical characteristics of MWCNTs [56]. Compared to MWCNTs with three or more layers, DWCNTs have a smaller diameter and fewer flaws. According to theory, the transport characteristics of specific DWCNTs exhibit a significant variation in conductance that corresponds to variations in the chirality of the inner and outer walls [57].

1.1.3.6 Multi-walled Carbon Nanotubes (MWCNTs)

The larger and more complex multi-walled carbon nanotube is made up of numerous single-walled tubes stacked one on top of the other. Depending on the quantity of graphene tubes, MWNTs are made up of numerous layers of graphene that have been rolled into tubes with sizes ranging from 2 to 50 nm. These tubes have an interlayer spacing of almost 0.34 nm [58]. MWCNTs are made of carbon atoms with thicker walls and are made up of numerous coaxial graphene cylinders that are spaced apart by a gap that is probably not much greater than the interlayer distance in graphite. The inner hollow cylinder's diameters are 1–8 nm, and the MWCNTs' outer cylinder diameters range from 2 to 24 nm. MWCNTs have relatively simpler synthesis process and demonstrate better dispersibility over SWCNTs owing to their longer diameter in the presence of multi-layer graphene sheets. Consequently, the SWCNT bundle tends to occupy a larger area compared to MWCNTs for a specific interconnected length. Moreover, MWCNTs provide better structural diversity and subsequent tunable dispersion and solubility during deposition. Apart from interconnect application, SWCNTs also show promise for energy storage and electronics applications. In essence, the large electrical and thermal conductivity, higher mechanical strength (15–20 times stronger than steel), with lighter weight of MWCNTs make them highly attractive materials for interconnect applications.

1.1.4 CONCLUSION AND FUTURE SCOPE

The fascinating characteristics of carbon nanomaterials have sparked interest in nano-electronics. Three in-plane electronic orbitals and an out-of-plane p_z (π) orbital are introduced by the sp^2 orbital hybridization, which is primarily responsible for strong C–C bonding and low-energy electron transport. The band structure of these species, which shows a linear E–k low-energy band structure with diminishing effective mass and almost symmetric but exceptionally high electron and hole mobility, makes clear the originality of these species. In addition to exceptional thermal conductivity, the tightly bound hexagonal carbon atoms maintain mechanical toughness even at high temperatures and huge current conditions. Single atomic layer carbon nanomaterials are extremely helpful in difficult situations because of how resistant they are to short channel effects due to their thickness.

In addition to interconnects made of carbon nanomaterials, other upcoming interconnect technologies include laser-induced optical interconnects (OIs) and radio frequency or wireless interconnects (RFIs). These devices can transmit information at the speed of light in both micro-circuits and communication networks along with high information rate. Finally, yet importantly, carbon nanomaterials are perfectly suited for electronic/interconnect applications and merit significant research attention to fulfill their potential.

REFERENCES

[1] Li, Hong, Chuan Xu, Navin Srivastava, and Kaustav Banerjee. "Carbon nanomaterials for next-generation interconnects and passives: Physics, status, and prospects." *IEEE Transactions on Electron Devices* 56, no. 9 (2009) 1799.
[2] Li, Hong, Chuan Xu, and Kaustav Banerjee. "Carbon nanomaterials: The ideal interconnect technology for next-generation ICs." *IEEE Design and Test of Computers* 27, no. 4 (2010) 20.
[3] Chatterjee, S., and A. Chen. "Biosens." *Bioelectron* 35 (2012) 302.
[4] Tiwari, A., N. Bahadursha, J. Palepu, S. Chakraborty, and S. Kanungo. "Comparative analysis of Boron, nitrogen, and phosphorous doping in monolayer of semi-metallic xenes (graphene, silicene, and germanene): A first principle calculation based approach." *Materials Science in Semiconductor Processing* 153 (2023) 107121.
[5] Wanekaya, A. K. "Applications of nanoscale carbon-based materials in heavy metal sensing and detection." *Analyst* 136 (2011) 4383.
[6] Lim, E. K. T., Kim, S. Paik, S. Haam, Y. M. Huh, and K. Lee. "Nanomaterials for theranostics: recent advances and future challenges." *Chemical Reviews* 115 (2015) 327.
[7] Ramaiyan K. P., and R. Mukundan. "Electrochemical sensors for air quality monitoring." *Electrochemical Society Interface* 28 (2019) 59.
[8] Acquah, S. F. A., A. V. Penkova, D. A. Markelov, A. S. Semisalova, B. E. Leonhardt, and J. M. Magi. "The beautiful molecule: 30 years of C60 and its derivatives." *ECS Journal of Solid State Science and Technology* 6 (2017) M3155.
[9] Avouris, P., Z. Chen, and V. Perebeinos. "Carbon-based electronics." *Nature Nanotechnology* 2 (2007) 605.
[10] Liu, Z., J. T. Robinson, S. M. Tabakman, K. Yang, and H. Dai. "Carbon materials for drug delivery & cancer therapy." *Materials Today* 14 (2011) 316.

[11] Mendes, R. G., A. Bachmatiuk, B. Buechner, G. Cuniberti, and M. H. Ruemmeli. "Carbon nanostructures as multi-functional drug delivery platforms." *Journal of Materials Chemistry B* 1 (2013) 401.

[12] Zaytseva, O., and G. Neumann. "Carbon nanomaterials: Production, impact on plant development, agricultural and environmental applications." *Chemical and Biological Technologies in Agriculture* 3 (2016) 17.

[13] Kamel, Samir, Mohamed El-Sakhawy, Badawi Anis, and Hebat-Allah S. Tohamy. "Graphene's structure, synthesis, and characterization: A brief review." *Egyptian Journal of Chemistry: Innovation in Chemistry* 62, no. 2 (2019) 593.

[14] Novoselov, K. S., A. K. Geim, S. V. Morozov, D. Jiang, Y. Zhang, S. V. Dubonos, I. V. Grigorieva, and A. Firsov. "Electric field effect in atomically thin carbon films." *Science* 306 (2004) 666.

[15] Novoselov, K. S., A. K. Geim, S. V. Morozov, D. Jiang, M. I. Katsnelson, I. V. Grigorieva, S. V. Dubonos, and A. A. Firsov. "Two-dimensional gas of massless Dirac fermions in graphene." *Nature* 438 (2005) 197.

[16] Geim, A. K. "Graphene: Status and prospects." *Science* 324 (2009) 1530.

[17] Lemme, Max C. "Current status of graphene transistors." *Solid State Phenomena* 156–158 (2010) 499.

[18] Hwang, Harold Y., Yoh Iwasa, Masashi Kawasaki, Bernhard Keimer, Naoto Nagaosa, and Yoshinori Tokura. "Emergent phenomena at oxide interfaces." *Nature Materials* 11 (2012) 103.

[19] Tiwari, A., J. Palepu, A. Choudhury, S. Bhattacharya, and S. Kanungo. "Theoretical analysis of the NH_3, NO, and NO_2 adsorption on boron-nitrogen and boron-phosphorous co-doped monolayer graphene: A comparative study." *FlatChem* 34 (2022) 100392.

[20] Kwon, Soon-Yong, Cristian V. Ciobanu, Vania Petrova, Vivek B. Shenoy, Javier Bareno, Vincent Gambin, Ivan Petrov, and Suneel Kodambaka. "Growth of semiconducting graphene on palladium." *Nano Letters* 9, no. 12 (2009) 3985.

[21] Sutter, Peter W., Jan-Ingo Flege, and Eli A. Sutter. "Epitaxial graphene on ruthenium." *Nature Materials* 7 (2008) 406.

[22] Coraux, Johann, Alpha T. N'Diaye, Carsten Busse, and Thomas Michely. "Structural coherency of graphene on Ir(111)." *Nano Letters* 8, no. 2 (2008) 565.

[23] Xuesong, Li, Weiwei Cai, Jinho An, Seyoung Kim, Junghyo Nah, Dongxing Yang, Richard Piner, Aruna Velamakanni, Inhwa Jung, Emanuel Tutuc, Sanjay K. Banerjee, Luigi Colombo, and Rodney S. Ruoff. "Large-area synthesis of high-quality and uniform graphene films on copper foils." *Science* 324 (2009) 1312.

[24] Mattevi, Cecilia, Hokwon Kima, and Manish Chhowalla. "A review of chemical vapour deposition of graphene on copper." *Journal of Material Chemistry* 21 (2011) 3324.

[25] Lee, Changgu, Xiaoding Wei, Jeffrey W. Kysar, and James Hone. "Measurement of the elastic properties and intrinsic strength of monolayer graphene." *Science* 321 (2008) 385.

[26] Monie, Sanjay. *Developments in Conductive Ink*. https://web.archive.org/web/20140414140145/http://industrial-printing.net/content/developments-conductive-inks?page=0,3

[27] Kawai, Shigeki, Soichiro Nakatsuka, Takuji Hatakeyama, Rémy Pawlak, Tobias Meier, John Tracey, Ernst Meyer, and Adam S. Foster. "Multiple heteroatom substitution to graphene nanoribbon." *Science Advances* 4, no. 4 (2018) eaar7181.

[28] Wakabayashi, Katsunori, Mitsutaka Fujita, Koichi Kusakabe, and Kyoka Nakada. "Magnetic structure of graphite ribbon." *Czechoslovak Journal of Physics* 46, no. 4 (1996) 1865.

[29] Saraswat, Vivek, Robert M. Jacobberger, and Michael S. Arnold. "Materials science challenges to graphene nanoribbon electronics." *ACS Nano* 15, no. 3 (2021) 3674.

[30] Wakabayashi, Katsunori, Mitsutaka Fujita, Hiroshi Ajiki, and Manfred Sigrist. "Electronic and magnetic properties of nanographite ribbons." *Physical Review B* 59, no. 12 (1999) 8271.

[31] Son, Young-Woo, Marvin L. Cohen, and Steven G. Louie. "Energy gaps in graphene nanoribbons." *Physical Review Letters* 97, no. 21 (2006) 216803.

[32] Burgos, Palacios, Juan José, Joaquín Fernández-Rossier, and Luis Brey Abalo. "Vacancy-induced magnetism in graphene and graphene ribbons." *Physical Review B* 77, no. 19 (2008) 195428.

[33] Son, Young-Woo, Marvin L. Cohen, and Steven G. Louie. "Energy gaps in graphene nanoribbons." *Physical Review Letters* 97, no. 21 (2006) 216803.

[34] Fernández-Rossier, J. "Hidden multiferroic order in graphene zigzag ribbons." *Physical Review B* 77 (2008) 075430.

[35] Awschalom, David D., Ryan Epstein, and Ronald Hanson. "The diamond age of spintronics." *Scientific American* 297, no. 4 (2007) 84.

[36] Chappert, C., and A. Fert. "Nguyen Van Dau F." *Nature Mater* 6 (2007) 813.

[37] Son, Young-Woo, Marvin L. Cohen, and Steven G. Louie. "Energy gaps in graphene nanoribbons." *Physical Review Letters* 97, no. 21 (2006) 216803.

[38] Dimiev, Ayrat, Wei Lu, Kyle Zeller, Benjamin Crowgey, Leo C. Kempel, and James M. Tour. "Low-loss, high-permittivity composites made from graphene nanoribbons." *ACS Applied Materials and Interfaces* 3, no. 12 (2011) 4657.

[39] Dimiev, Ayrat, Dante Zakhidov, Bostjan Genorio, Korede Oladimeji, Benjamin Crowgey, Leo Kempel, Edward J. Rothwell, and James M. Tour. "Permittivity of dielectric composite materials comprising graphene nanoribbons: The effect of nanostructure." *ACS Applied Materials and Interfaces* 5, no. 15 (2013) 7567.

[40] Abukari, Sulemana S., Samuel Y. Mensah, Musah Rabiu, Kofi W. Adu, Natalia G. Mensah, Anthony Twum, Alfred Owusu, Kwadwo A. Dompreh, Patrick Mensah-Amoah, and Matthew Amekpewu. "High-frequency electric field induced nonlinear electron transport in chiral carbon nanotubes." *World Journal of Condensed Matter Physics*, 5 (2015) 294.

[41] Oberlin, Agnes, M. Endo, and T. Koyama. "Filamentous growth of carbon through benzene decomposition." *Journal of Crystal Growth* 32, no. 3 (1976) 335.

[42] Iijima, Sumio. "Helical microtubules of graphitic carbon." *Nature* 354, no. 6348 (1991) 56.

[43] Saifuddin, N., A. Z. Raziah, and A. R. Junizah. "Carbon nanotubes: A review on structure and their interaction with proteins." *Journal of Chemistry* 2013 (2013) 676815.

[44] Thiruvengadam, Muthu, Govindasamy Rajakumar, Venkata Swetha, Mohammad Azam Ansari, Saad Alghamdi, Mazen Almehmadi, Mustafa Halawi et al. "Recent insights and multifactorial applications of carbon nanotubes." *Micromachines* 12, no. 12 (2021) 1502.

[45] Zhou, Tao, Yutao Niu, Zhi Li, Huifang Li, Zhenzhong Yong, Kunjie Wu, Yongyi Zhang, and Qingwen Li. "The synergetic relationship between the length and orientation of carbon nanotubes in direct spinning of high-strength carbon nanotube fibers." *Materials and Design* 203 (2021) 109557.

[46] Singh, Ekta, Richa Srivastava, Utkarsh Kumar, and Anamika D. Katheria. "Carbon nanotube: A review on introduction, fabrication techniques and optical applications." *Journal of Nanoscience & Nanotechnology Research* 4 (2017) 120.

[47] Tîlmaciu, C. M., and M. C. Morris. "Carbon nanotube biosensors." *Frontiers in Chemistry* 3 (2015) 1.

[48] Han, Zhidong, and Alberto Fina. "Thermal conductivity of carbon nanotubes and their polymer nanocomposites: A review." *Progress in Polymer Science* 36, no. 7 (2011) 914.

[49] Baxendale, M. "The physics and applications of carbon nanotubes." *Journal of Materials Science: Materials in Electronics* 14, no. 10 (2003) 657.

[50] Ruland, W., A. K. Schaper, Haoqing Hou, and Andreas Greiner. "Multi-wall carbon nanotubes with uniform chirality: Evidence for scroll structures." *Carbon* 41, no. 3 (2003) 423.

[51] Yang, Qiuyue, Emily P. Nguyen, Cecilia de Carvalho Castro Silva, Giulio Rosati, and Arben Merkoçi. "Signal enhancement strategies." In: *Wearable Physical, Chemical and Biological Sensors*, pp. 123–168 (Elsevier, Amsterdam, 2022).

[52] Abdalla, S., F. Al-Marzouki, Ahmed A. Al-Ghamdi, and A. Abdel-Daiem. "Different technical applications of carbon nanotubes." *Nanoscale Research Letters* 10, no. 1 (2015) 1.

[53] Ganesh. E. N. "Single walled and multi walled carbon nanotube structure, synthesis and applications." *International Journal of Innovative Technology and Exploring Engineering* 2, no. 4 (2013) 311.

[54] Goddard III, William A., Donald Brenner, Sergey Edward Lyshevski, and Gerald J. Iafrate. *Handbook of Nanoscience, Engineering, and Technology* (CRC Press, London, 2002).

[55] Dresselhaus, G., Mildred S. Dresselhaus, and Riichiro Saito. *Physical Properties of Carbon Nanotubes* (World Scientific, Singapore, 1998).

[56] Hirlekar, R., M. Yamagar, H. Garse, M. Vij, and V. Kadam. Nanotubes and Its Applications: A Review. *Asian Journal of Pharmaceutical and Clinical Research* 2, no. 4 (2009) 17.

[57] Eslami, Majid, Vahid Vahabi, and Ali Ahmadi Peyghan. "Sensing properties of BN nanotube toward carcinogenic 4-chloroaniline: A computational study." *Physica E: Low-dimensional Systems and Nanostructures* 76 (2016) 6.

[58] Kazaoui, S., N. Minami, R. Jacquemin, H. Kataura, and Y. Achiba. "Amphoteric doping of single-wall carbon-nanotube thin films as probed by optical absorption spectroscopy." *Physical Review B* 60, no. 19 (1999) 13339.

[59] Georgakilas, V., Perman, J. A., Tucek, J., & Zboril, R. "Broad family of carbon nanoallotropes: classification, chemistry, and applications of fullerenes, carbon dots, nanotubes, graphene, nanodiamonds, and combined superstructures." Chemical reviews, 2015, 4744–4822.

[60] Afreen, S., Omar, R. A., Talreja, N., Chauhan, D., & Ashfaq, M. "Carbon-Based Nanostructured Materials for Energy and Environmental Remediation Applications: The New Era of Environmental Microbiology and Nanobiotechnology." *Approaches in bioremediation: The new era of environmental microbiology and nanobiotechnology* (2018) 369–392.

2 Interconnect Modeling Using Graphene Nanoribbon (GNR)

Wen-Sheng Zhao

2.1 INTRODUCTION

With the advances of fabrication technology, the integrated circuit (IC) feature size is continuously shrinking [1]. Unlike transistors, miniaturization of interconnects does not enhance their performance [2]. The scaling down of interconnect dimensions increases the probability of scattering of electrons at surface and grain boundary, thereby dramatically increasing the effective resistivity [3,4]. The increase in the interconnect resistance would result in serious performance degradation and reliability problems [5–7]. The interconnect delay has become larger than the transistor delay, and the interconnect consumes more than half of the total power dissipated in the current microprocessor. As the influence of interconnect on the IC performance is becoming increasingly prominent, the IC technology has evolved from a transistor-centric era into an interconnect-centric era [8].

To solve the interconnect issue, various alternative conductor materials have been proposed. Graphene, which was first discovered in 2004 [9], has also been used to build interconnect due to its extraordinary physical properties such as long mean free path (MFP), high thermal conductivity, and strong tensile strength [10–14]. In comparison with carbon nanotube (CNT), graphene is more compatible with conventional CMOS process. It was experimentally demonstrated that graphene nanoribbon (GNR) has an ultrahigh breakdown current density larger than 10^8 A/cm^2, which is critical for future nanoscale ICs [15]. To enable large-scale fabrication, the chemical vapor deposition (CVD) method was utilized to produce GNRs, whose peak current density achieved a high value of 2×10^9 A/cm^2 [16]. By intercalation-doping techniques, the electronic conductance of GNR interconnect was dramatically improved [17]. Furthermore, the GNR interconnect was combined with the CNT vias to form all-carbon three-dimensional (3D) interconnects [18,19]. It is anticipated that such all-carbon 3D interconnects could provide superior electrical and thermal performances to conventional copper wires [20,21].

This chapter will focus on the circuit modeling of GNR interconnects. We start from some basic electronic properties of GNRs in Section 2.2. Then, a set of equations is presented for fast calculation of the number of effective conducting channels of GNR. The transmission line model of monolayer GNR interconnect is established

DOI: 10.1201/9781003331650-2

in Section 2.3, with quantum resistance, kinetic inductance, and quantum capacitance extracted analytically. In Section 2.4, the influence of contact type on the conductance of multilayer GNR interconnect is investigated. Then, the vertical GNR interconnect scheme is proposed and studied in Section 2.5. Finally, some conclusions are drawn in Section 2.6.

2.2   ELECTRONIC PROPERTY OF GNR

Figure 2.1 shows the atomic structure of the graphene layer. In the figure, a denotes the spacing between neighboring carbon atoms.

GNR, one-dimensional (1D) form of graphene, can be cut from the graphene layer, and its charity is defined as [22]

$$C = na_1 + ma_2 \qquad (2.1)$$

where n and m are chirality indexes, and a_1 and a_2 are the lattice vectors of graphene. The GNR would be zigzag type for $n = m$, and otherwise, it would be the armchair type. Figure 2.2 shows the first Brillouin zone of the graphene layer, and the reciprocal lattice vectors are given as

$$b_1 = \frac{2\pi}{3a}\left(1, \sqrt{3}\right) \qquad (2.2)$$

$$b_2 = \frac{2\pi}{3a}\left(1, -\sqrt{3}\right) \qquad (2.3)$$

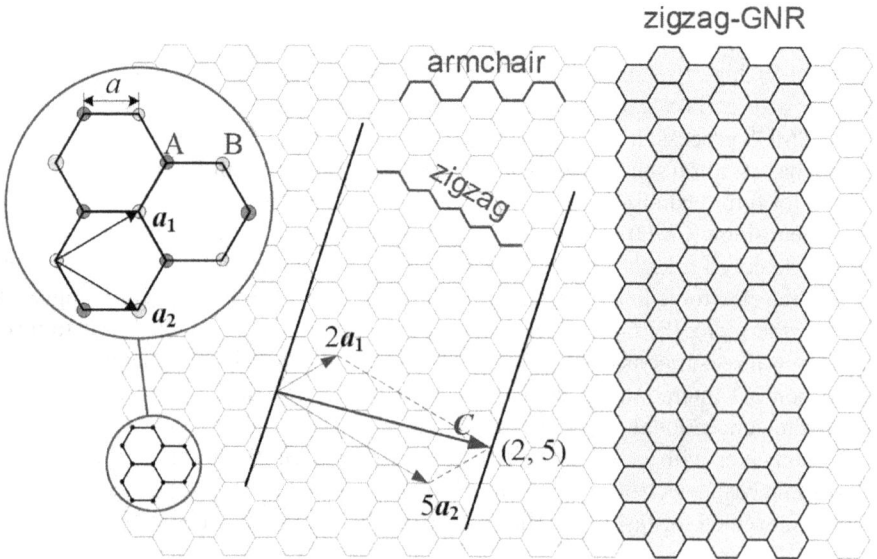

FIGURE 2.1   Atomic structure of graphene layer.

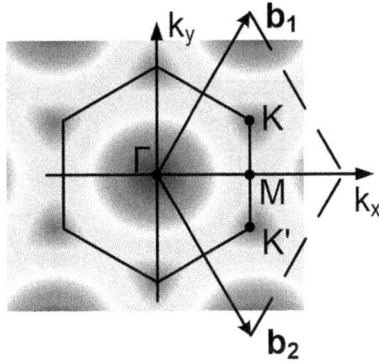

FIGURE 2.2 Brillouin zone of graphene.

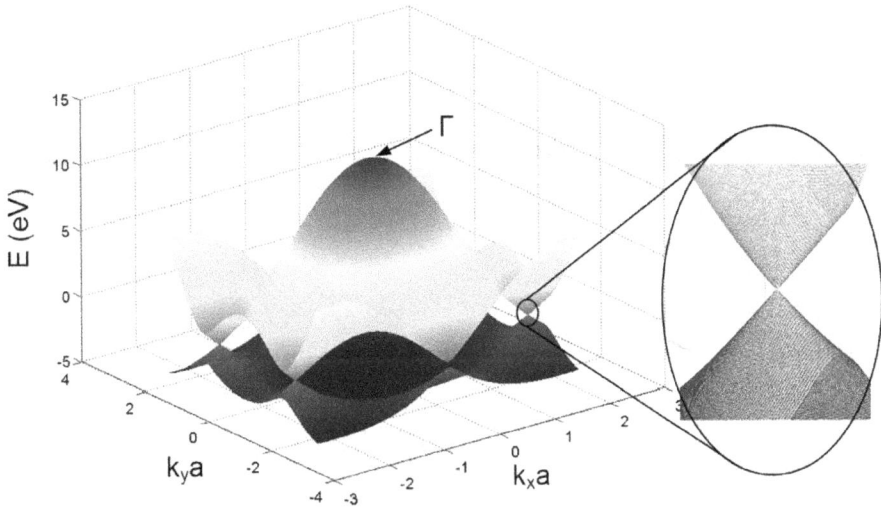

FIGURE 2.3 Band structure of graphene.

The unique physical property of graphene is determined by its energy band structure, which can be obtained according to tight-bound approximation:

$$E_{\pm}(k) = \pm\gamma\sqrt{3 + f(k)} - \gamma f(k) \qquad (2.4)$$

$$f(k) = 4\cos\left(\frac{3}{2}k_{xa}\right)\cos\left(\frac{\sqrt{3}}{2}k_y a\right) + 2\cos\left(\sqrt{3}k_y a\right) \qquad (2.5)$$

where $k = k_x\hat{x} + k_y\hat{y}$ is the wave vector of the electron wave function, and γ is the transition energy value [23]. As shown in Figure 2.3, the energy band shows a linear relationship near the Dirac point, i.e., $E = v_F\hbar k$, where v_F is the Fermi velocity and $\hbar$

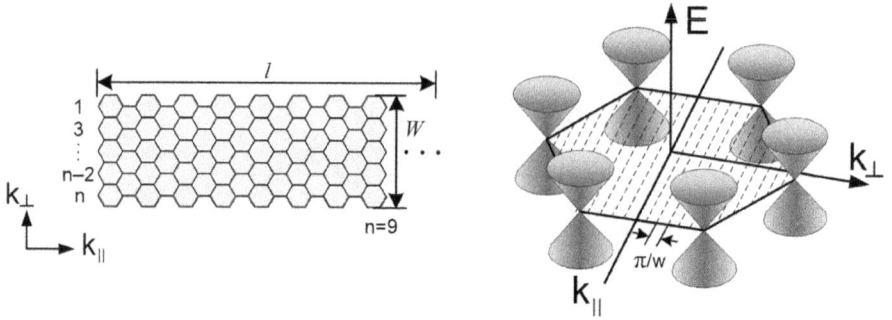

FIGURE 2.4   Schematic of armchair GNR and its energy band structure.

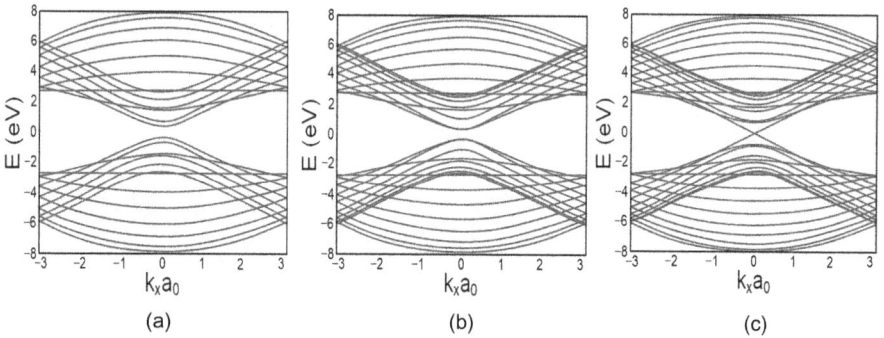

FIGURE 2.5   Band diagram of armchair GNRs: (a) $n = 15$, (b) $n = 16$, and (c) $n = 17$.

is the reduced Planck constant. Under such circumstance, the energy band can be approximated as

$$E_{\pm}(k) = \pm\gamma\sqrt{3 + f(k)} \tag{2.6}$$

As aforementioned, the GNR is cut from the graphene layer. As shown in Figure 2.4, an armchair GNR is used as interconnect, and its lattice constant is $a_0 = 3a$. Due to the finite width of GNR, the wave vector can be obtained as $k_{\perp} = i\pi/W$, where i is an integer and W is the GNR width. By substituting $k_{\perp}$ into equation (2.6), the band diagram of armchair GNR can be obtained, as shown in Figure 2.5. It is found that when the number of carbon atoms $n = 3q + 2$, where q is a positive integer, the armchair GNR would be metallic, and for $n = 3q$ or $3q + 1$, the armchair GNR is semiconducting. Unlike the armchair GNR, the zigzag GNR has a lattice constant of $a_0 = \sqrt{3}a$, and it is always metallic.

Considering the ballistic transport, the GNR conductance can be calculated as $G = G_0 N_{ch}$, where $G_0 = 2e^2/h$ is the quantum conductance, e is the electron charge, h is the Planck constant, and N_{ch} is the number of effective conducting channels. The electrons in GNR obey the Fermi-Dirac statistics

$$f(E) = \frac{1}{1 + e^{\frac{E - E_F}{k_B T}}} \qquad (2.7)$$

where E_F is the Fermi energy, k_B is the Boltzmann constant, and T is the temperature. According to the Fermi-Dirac distribution function, N_{ch} can be calculated by [24]

$$N_{ch} = \sum_{i=1}^{n_c} \frac{1}{1 + e^{\frac{E_i - E_F}{k_B T}}} + \sum_{i=1}^{n_v} \frac{1}{1 + e^{\frac{E_F - E_i}{k_B T}}} \qquad (2.8)$$

According to [25], the energy band of GNR can be approximated by

$$N_{ch} = \sum_{i=1}^{n_c} \frac{1}{1 + e^{\frac{E_i - E_F}{k_B T}}} + \sum_{i=1}^{n_v} \frac{1}{1 + e^{\frac{E_F - E_i}{k_B T}}} \qquad (2.9)$$

By substituting the quantized transverse wave vector into equation (2.8), the conduction energy band of armchair GNR and zigzag GNR can be given as [26]

$$E_i = \begin{cases} nE|i|, \text{ metallic} \\ nE\left|i + \dfrac{1}{3}\right|, \text{ semiconducting} \end{cases} \quad \text{armchair GNR} \qquad (2.10)$$

$$E_i = \begin{cases} 0, i = 0 \\ nE\left(|i| + \dfrac{1}{2}\right), i \neq 0 \end{cases} \quad \text{zigzag GNR} \qquad (2.11)$$

where $nE = h v_F / (2W)$. Actually, the bandgap would be introduced into the zigzag GNR due to the influence of electron degeneracy, and the energy band becomes [27]

$$E_i = \begin{cases} \dfrac{0.467}{W + 1.5}, i = 0 \\ nE\left(|i| + \dfrac{1}{2}\right), i \neq 0 \end{cases} \quad \text{zigzag GNR} \qquad (2.12)$$

As shown in Figure 2.6, the bandgap of zigzag GNR only has an impact for $E_F = 0$. N_{ch} increases with increasing E_F. Similar to the CNT, when the width exceeds a certain value, the semiconducting armchair GNR could also be used as conductor. This is because the difference between minimum conduction energy and Fermi energy would be smaller than thermal energy $k_B T$, and the electron transition may appear due to thermal excitation. Under such circumstances, both the metallic and semiconducting armchair GNRs possess larger N_{ch} than the zigzag GNR. Therefore, the metallic armchair GNR is used for interconnect modeling in the following. The number of effective conducting channels of the metallic armchair GNR vs the Fermi energy and width is plotted in Figure 2.7. It is found that N_{ch} increases linearly for $E_F > 0.1 \text{eV}$ and $W > 10 \text{nm}$, and it can be approximated as [28]

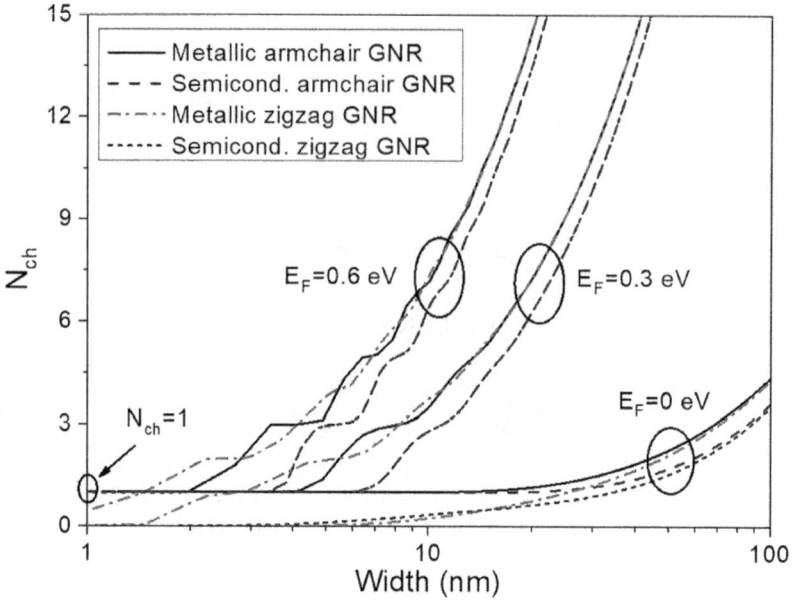

FIGURE 2.6   Number of effective conducting channels of GNR vs width.

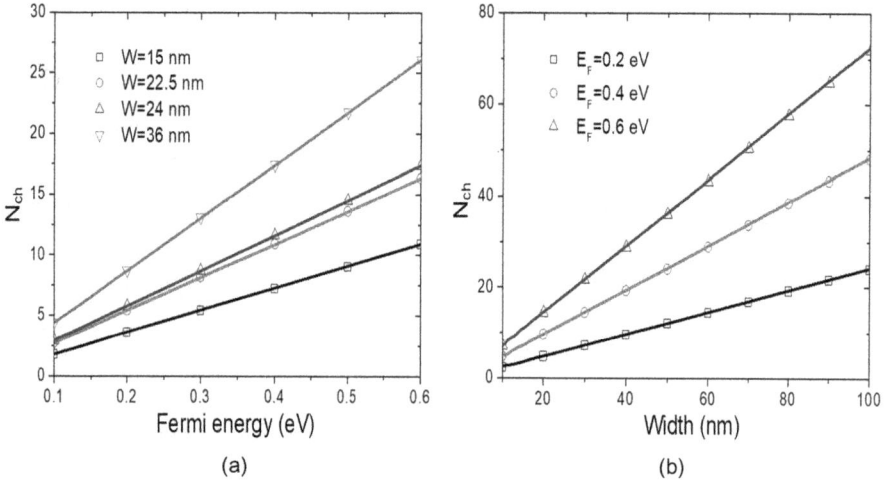

FIGURE 2.7   Number of effective conducting channels of metallic armchair GNR vs (a) Fermi energy and (b) width.

$$N_{ch} = \begin{cases} a_1W + a_2, \ E_F = 0 \\ a_3WE_F - a_4, \ E_F \geq 0.1\text{eV} \end{cases} \qquad (2.13)$$

where $a_1 = 0.029\,\text{nm}^{-1}$, $a_2 = 0.057$, $a_3 = 1.20\ \text{eV}^{-1}\,\text{nm}^{-1}$, and $a_4 = 0.82$. In the figure, the solid lines are calculated from equation (2.13), and the symbols are obtained numerically. As the MFP of GNR may be affected by the edge scattering, the GNR resistance for non-specular scattering at edges still needs to be extracted numerically.

2.3   MODELING OF MONOLAYER GNR INTERCONNECT

Figure 2.8 shows the structure of monolayer GNR interconnect above the ground plane together with the equivalent circuit model. For ballistic transport, the GNR resistance can be calculated as quantum resistance [29]

$$R_Q = \frac{h}{2e^2} \frac{1}{N_{ch}} \approx \frac{12.9}{N_{ch}} \, k\Omega \tag{2.14}$$

As the interconnect length exceeds the MFP, the GNR should be modeled as distributed circuit model, with the scattering resistance, kinetic inductance, and quantum capacitance [30] as

$$R_S = \frac{h}{4e^2} \frac{1}{\lambda} \tag{2.15}$$

$$L_K = \frac{h}{8e^2 v_F} \approx 4 \, nH/\mu m \tag{2.16}$$

$$C_Q = \frac{8e^2}{hv_F} \approx 100 \, aF/\mu m \tag{2.17}$$

where λ represents the effective MFP of the GNR. The electrostatic capacitance and magnetic inductance of monolayer GNR interconnect are given by [31]

$$C_E = \varepsilon M \left(\tanh\left(\frac{\pi W}{4d} \right) \right) \tag{2.18}$$

$$L_M = \mu \varepsilon / C_E \tag{2.19}$$

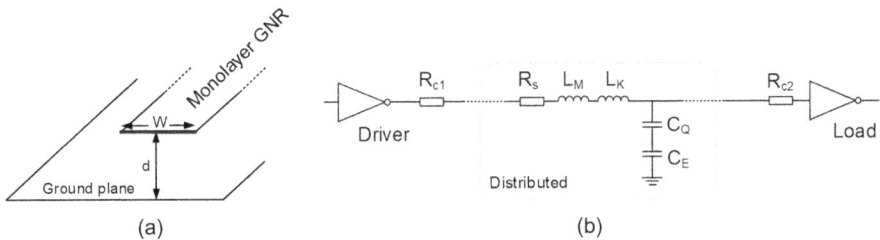

FIGURE 2.8   (a) Schematic of monolayer GNR interconnect and its (b) circuit model.

where ε and μ denote the permittivity and permeability of the surrounding dielectric, and d is the distance between the GNR and ground plane. The function $M(k)$ is given by

$$M(k) = \begin{cases} \dfrac{2\pi}{\ln\left(2\dfrac{1+\sqrt{1-k^2}}{1-\sqrt{1-k^2}}\right)}, & 0 \le k \le \dfrac{1}{\sqrt{2}} \\[4ex] \dfrac{2}{\pi}\ln\left(2\dfrac{1+\sqrt{k}}{1-\sqrt{k}}\right), & \dfrac{1}{\sqrt{2}} \le k \le 1 \end{cases} \tag{2.20}$$

It is evident that the GNR resistance highly depends on the effective MFP, which is about 1 μm for suspended graphene layer [32]. However, the effective MFP degrades when the graphene is placed on the substrate. The MFP is also affected by edge scattering, and it could be calculated by

$$\lambda_i = \frac{1}{1-p}\frac{v_\perp}{v_\parallel}W = \frac{W}{1-p}\sqrt{\left(\frac{E_F}{inE}\right)^2 - 1} \tag{2.21}$$

where $v_\perp$ and $v_\parallel$ denote the longitudinal and transverse electron velocities, respectively, and p is the specularity constant. The specularity constants of 0 and 0.8 have been reported in experiments [33]. The effective MFP of each subband can be calculated by

$$\lambda_{i,eff} = \left(\frac{1}{\lambda_d} + \frac{1}{\lambda_i}\right)^{-1} \tag{2.22}$$

where λ_d is the MFP corresponding to the defect scattering. Considering the edge scattering, the total resistance of the monolayer GNR interconnect could be calculated by [26]

$$R_S = \frac{h}{2e^2}\left(\sum_i \frac{1}{1+\dfrac{1}{\lambda_{i,eff}}}\right)^{-1} \tag{2.23}$$

Here, the imperfect contact resistance is neglected as it highly depends on the fabrication process, and therefore, the contact resistance R_c is equivalent to R_Q. Figure 2.9 shows the per-unit-length (p.u.l.) resistance of the monolayer GNR interconnect. It is evident that the edge scattering plays a significant impact on the GNR resistance for $W > 10$ nm. In certain conditions, the GNR resistance is close to the resistance of single-walled CNT bundle interconnect. In particular, when E_F is larger than 0.4 eV, the GNR interconnect becomes superior to the ideal single-walled CNT bundle interconnect. However, even in the ideal condition, the resistance of monolayer GNR

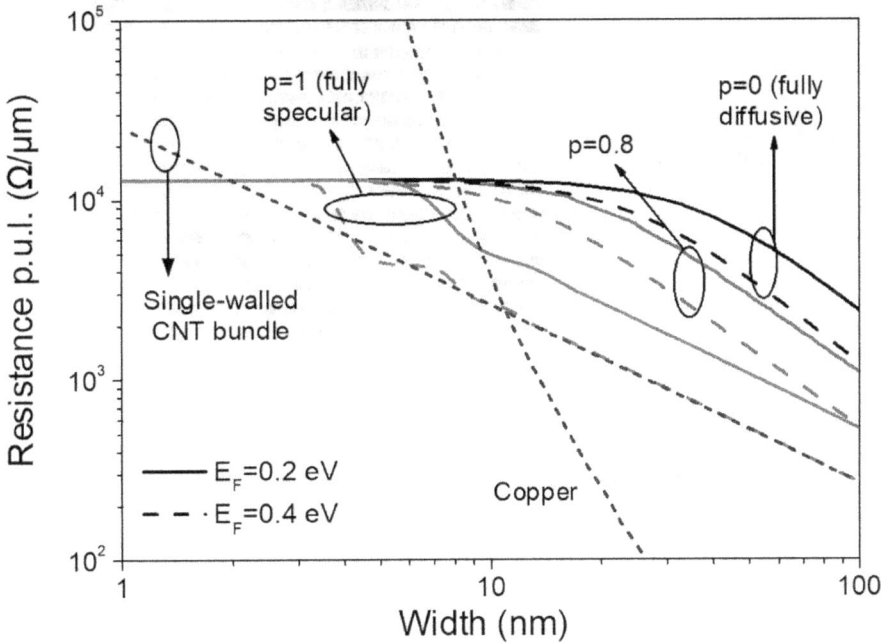

FIGURE 2.9   Distributed resistance of the monolayer GNR interconnect.

interconnect is still larger than that of conventional copper wire. Therefore, it is necessary to employ multilayer GNR as interconnect conductor material.

2.4   MODELING OF MULTILAYER GNR INTERCONNECT

To reduce resistive loss, the multilayer GNR is investigated for interconnect applications [34–36], as shown in Figure 2.10. The interconnect width and thickness are denoted as W and T, respectively, and the layer number is $N = 1 + \mathrm{inter}(T/\delta)$. δ is the distance between adjacent graphene layers, and it is usually 0.34 nm. It is worth noting that in most modeling studies, the side contacts are assumed, and all the graphene layers are connected to the electrodes at two ends. However, the top contacts are usually used in real-world applications, i.e., only the top graphene layer is coupled to the electrodes, as shown in Figure 2.11. According to [37], the top-contacted multilayer GNR interconnect can be modeled as a distributed resistance network, with the interlayer perpendicular resistance R_{perp} treated appropriately. The analyzed results imply that with increasing interconnect length, the interlayer perpendicular resistance decreases dramatically, while the in-layer resistance R_{layer} increases [38]. Therefore, when the length exceeds a certain value, the influence of the contact type could be neglected [28]. That is, for intermediate-level and global-level interconnects, both the side-contacted and top-contacted multilayer GNR interconnects can be modeled as multi-conductor transmission line circuit model [39,40], as shown in Figure 2.12.

$\delta = 0.34$ nm

T

ε

d

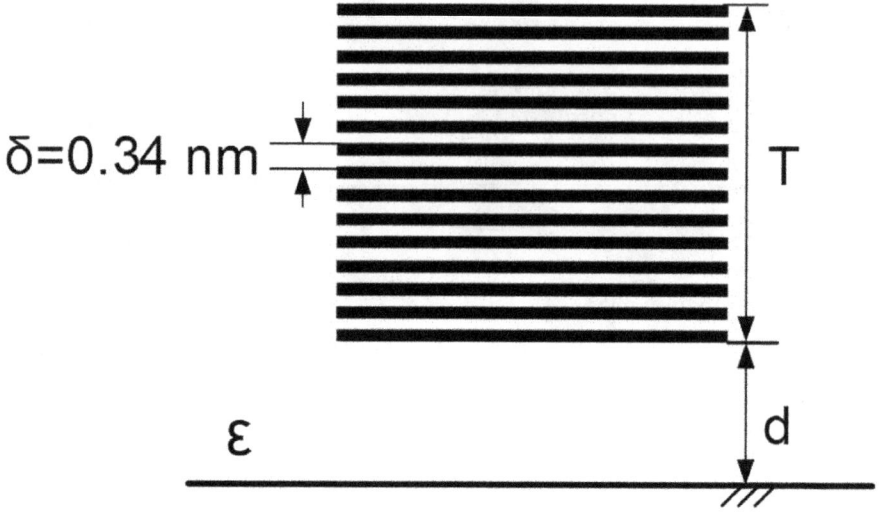

FIGURE 2.10   Cross-sectional view of multilayer GNR interconnect.

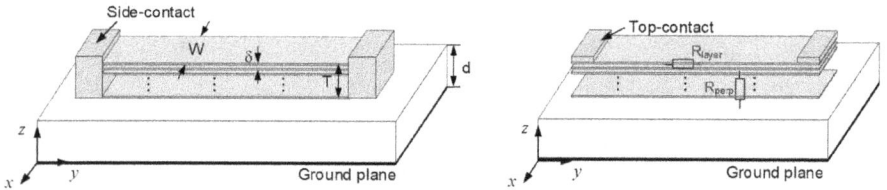

FIGURE 2.11   Schematics of side-contacted and top-contacted multilayer GNR interconnects.

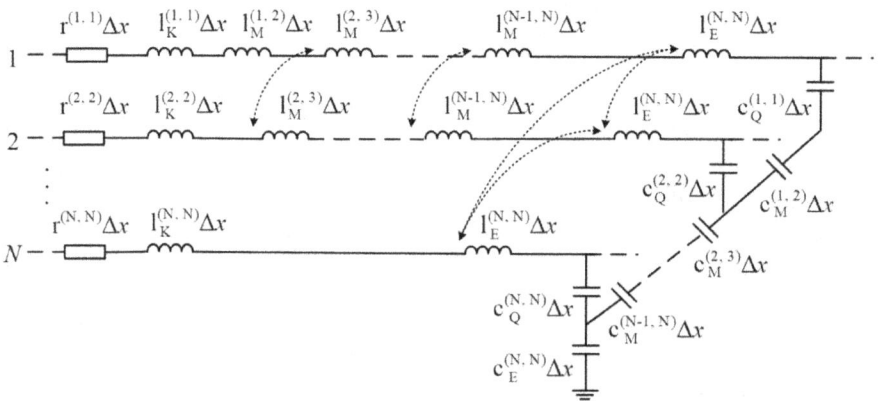

FIGURE 2.12   Multi-conductor transmission line circuit model of the multilayer GNR interconnect.

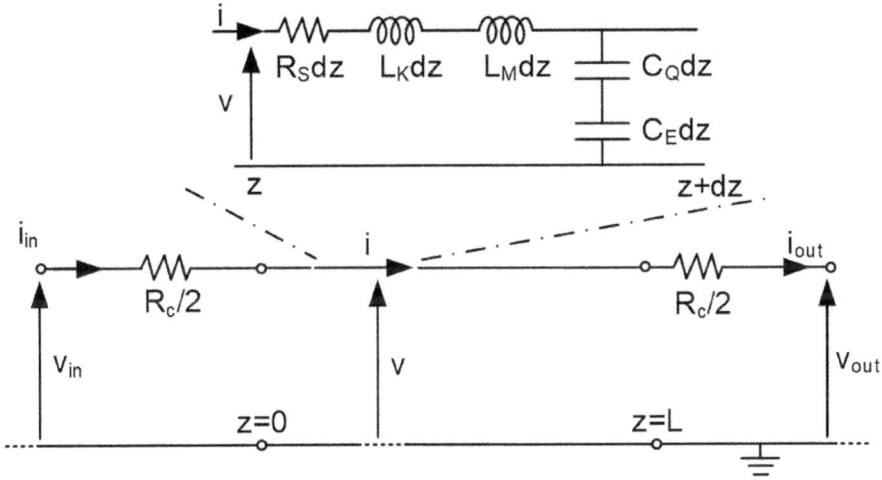

FIGURE 2.13   ESC transmission line circuit model of multilayer GNR interconnect.

To facilitate interconnect modeling, the multi-conductor model of multilayer GNR interconnect can be simplified as an equivalent single-conductor (ESC) transmission line circuit model, as shown in Figure 2.13 [41]. The p.u.l. equivalent quantum capacitance C_Q could be obtained through iteration [42]

$$c_{rec,1} = C_Q^{(1,1)} \tag{2.24}$$

$$c_{rec,i} = \left(\frac{1}{C_{rec,i-1}} + \frac{1}{C_m^{(i-1,i)}} \right)^{-1} + C_Q^{(i,i)} \tag{2.25}$$

where $C_m^{(i-1,i)} = \varepsilon_0 W/\delta$ and $C_Q^{(i,i)} = 4e^2 N_{ch}/(h v_F)$. Similarly, the equivalent kinetic inductance of multilayer GNR interconnect can be obtained through iteration. It is found that when the layer number exceeds a certain value (e.g., 10), the equivalent quantum capacitance tends to be stable, while the equivalent kinetic inductance decreases with increasing layer number, as shown in Figure 2.14. The ESC quantum capacitance and ESC kinetic inductance can be calculated by the following analytical expressions:

$$C_Q = \frac{2e^2 N_{ch}}{h v_F} + 2\sqrt{\left(\frac{e^2 N_{ch}}{h v_F} \right)^2 + \frac{e^2 N_{ch}}{h v_F} \cdot C_E} \tag{2.26}$$

$$L_K = \frac{h}{4e^2 v_F N} \cdot \frac{1}{N_{ch}} \tag{2.27}$$

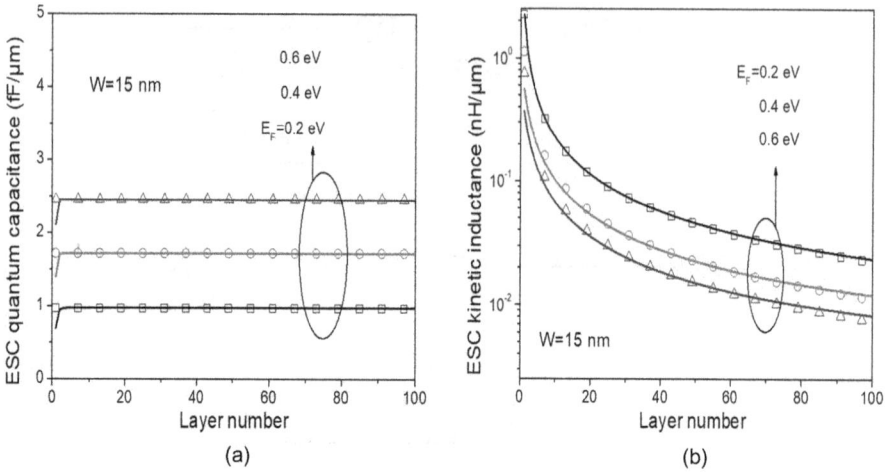

FIGURE 2.14   (a) ESC quantum capacitance and (b) ESC kinetic inductance of multilayer GNR interconnect [28].

In the figure, the solid lines are obtained by the above simplified expressions, while the symbols are obtained using recursive procedure. It is evident that two results agree well with each other. In general, the aspect ratio of on-chip interconnect is in the region of 2 and 3, and therefore, the above expressions are suitable for calculating the quantum capacitance and kinetic inductance of multilayer GNR interconnect. The total scattering resistance of multilayer GNR interconnect is calculated by

$$R_S = \frac{h}{2e^2} \frac{1}{NN_{ch}} \left(1 + \frac{l}{\lambda_d}\right) \tag{2.28}$$

For $0 \le p < 1$, the total scattering resistance of multilayer GNR interconnect is given by

$$R_S = \frac{h}{2e^2} \left[\sum_i \left(1 + \frac{l}{\lambda_{i,eff}}\right)^{-1}\right]^{-1} \tag{2.29}$$

Figure 2.15 shows the effective resistivities of multilayer GNR, multi-walled CNT, single-walled CNT bundle, and conventional copper interconnects. It is evident that the resistivity of multilayer GNR interconnect decreases with increasing Fermi energy, and is slightly larger than ideal single-walled CNT bundle interconnect. Based on the ESC transmission line circuit model, the 50% time delay of multilayer GNR interconnect can be obtained by [43]

$$\tau = \left(1.48\zeta + e^{-2.9\zeta^{1.35}}\right)\sqrt{lL(cL + C_L)} \tag{2.30}$$

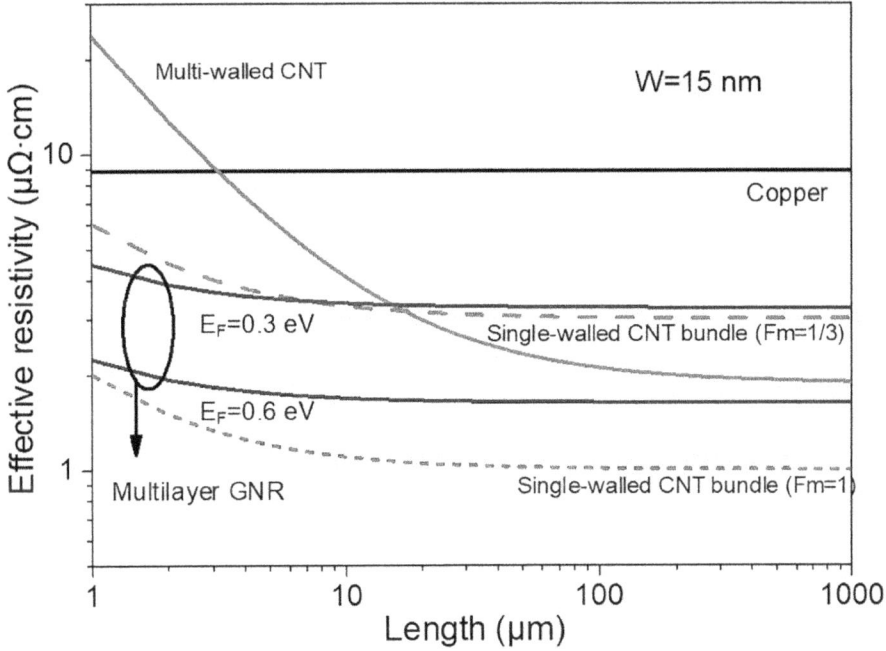

FIGURE 2.15   Resistivity of multilayer GNR interconnect [28].

where

$$\zeta = \frac{rL(cL + 2C_L) + 2(R_c + R_d)(cL + C_L)}{4\sqrt{lL(cL + C_L)}} \tag{2.31}$$

r, l, and c are the p.u.l. resistance, inductance, and capacitance of the interconnect, C_L is the load capacitance, and R_d is the driver resistance. The time delay ratio between copper wire and multilayer GNR interconnect is plotted in Figure 2.16 for intermediate level and global level. It is evident that the multilayer GNR interconnect could provide superior performance to conventional copper wire, and its advantage becomes more evident for advanced technology node. Moreover, the specularity constant p has a significant impact on the electrical performance. That is, the time delay of multilayer GNR interconnect can be dramatically reduced by increasing p.

2.5   MODELING OF VERTICAL GNR INTERCONNECT

Although the multilayer GNR interconnect could provide superior performance to conventional copper wire, its anisotropic property may result in serious problems. For example, the in-plane thermal conductivity of graphene layer could achieve more than 1,750 W/m K, but the out-plane thermal conductivity is only about 10 W/m·K [44]. It is evident that the low out-plane thermal conductivity of horizontal GNR interconnects would hinder heat dissipation in ICs. To resolve this issue, the vertical GNR interconnects are proposed and investigated for their electrical and thermal

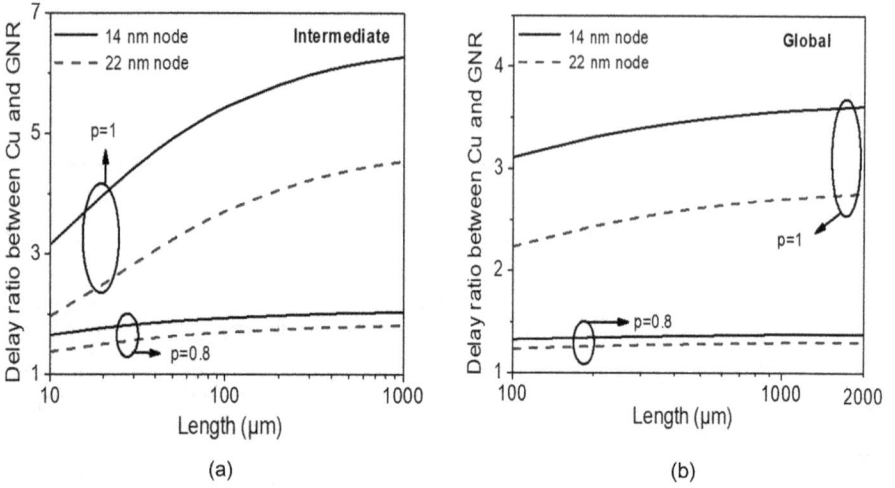

FIGURE 2.16   Delay ratio between copper and GNR interconnects at (a) intermediate and (b) global levels [28].

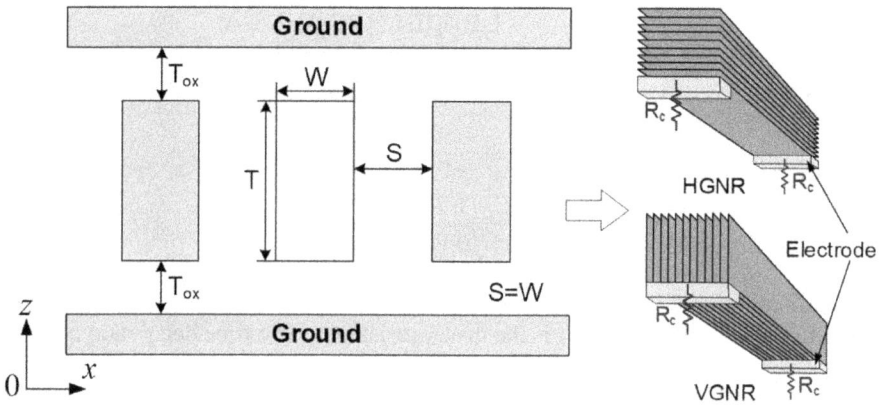

FIGURE 2.17   Schematic of horizontal GNR and vertical GNR interconnects.

performance [45,46]. As shown in Figure 2.17, each graphene layer in the vertical GNR interconnect could be effectively coupled to the electrodes. In the figure, S denotes the spacing between two adjacent interconnects, and T_{ox} is the thickness of inter-metal dielectric (IMD) layer.

The scattering resistance of the vertical GNR interconnect can be calculated as $R_S = R_{layer}/N$, where N is the layer number and R_{layer} is the resistance of each graphene layer, i.e.,

$$R_{layer} = \frac{h}{2e^2}\left[N_{ch,0}\left(1+\frac{l}{\lambda_d}\right)^{-1} + \sum_{i=1}^{M} N_{ch,i}\left(1+\frac{l}{\lambda_d}+\frac{l}{\lambda_i}\right)^{-1}\right]^{-1} \quad (2.32)$$

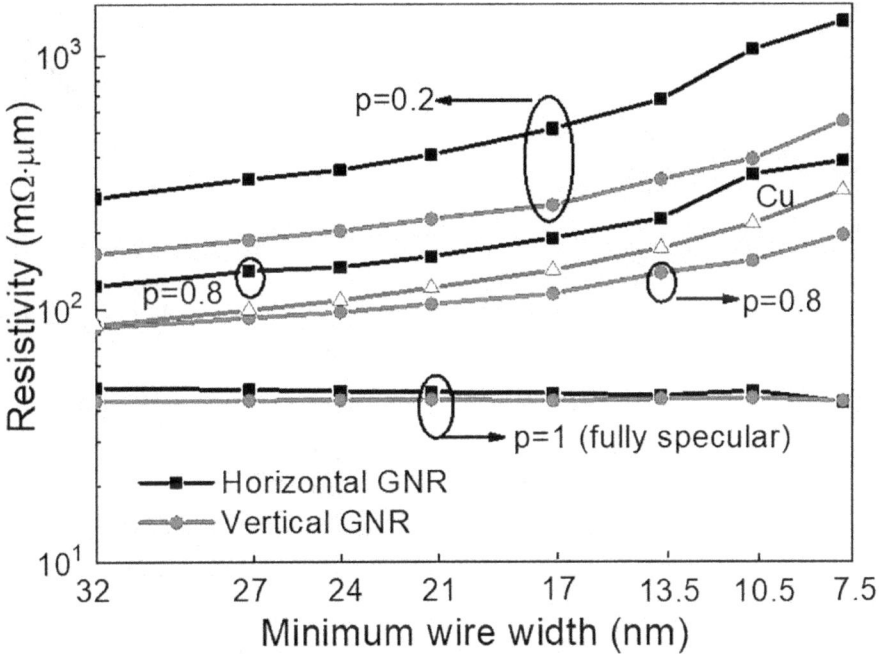

FIGURE 2.18   Effective resistivities of horizontal GNR and vertical GNR interconnects [46].

Here, the average MFP corresponding to the defect scattering is set as 419 nm according to [30], $M = \text{Inter}\left(E_F/nE\right)$, and $nE = 2\,\text{eV nm}/T$ is the distance between subbands [26]. For fully specular GNR interconnects, the resistance can be simplified as $R_{layer} = h/\left(2e^2 N_{ch}\right)\cdot\left(1 + l/\lambda_d\right)$ [28]. As the kinetic inductance of GNR is much larger than the magnetic inductance, the magnetic inductance can be neglected. Similarly, the quantum capacitance is negligible as it is much larger than the electrostatic one.

Figure 2.18 shows the effective resistivity of the horizontal GNR and vertical GNR interconnects at different technology nodes. In the figure, the interconnect length is 400 μm, Fermi energy is 0.2 eV, and the other geometrical parameters are adopted from ITRS prediction [47]. It is evident that the vertical GNR interconnect possesses smaller resistivity than the horizontal GNR interconnect. For fully specular, the effective resistivities of both horizontal GNR and vertical GNR interconnects remain almost unchanged with the advanced technology node. This is because for fully specular GNRs, N_{ch} is proportional to W, and the effective resistivity $\rho = R_Q \lambda_d WT/(NN_{ch})$ is irrespective of the interconnect width. However, for nonideal GNRs, the effective resistivities of both horizontal GNR and vertical GNR interconnects increase with decreasing width.

Based on the driver-interconnect-load system, the propagation delay of the vertical GNR interconnect is extracted using equation (4.30) and plotted in Figure 2.19. It is evident that the increase in specularity constant would improve the signal transmission performance of GNR interconnects, and narrow the gap between the horizontal

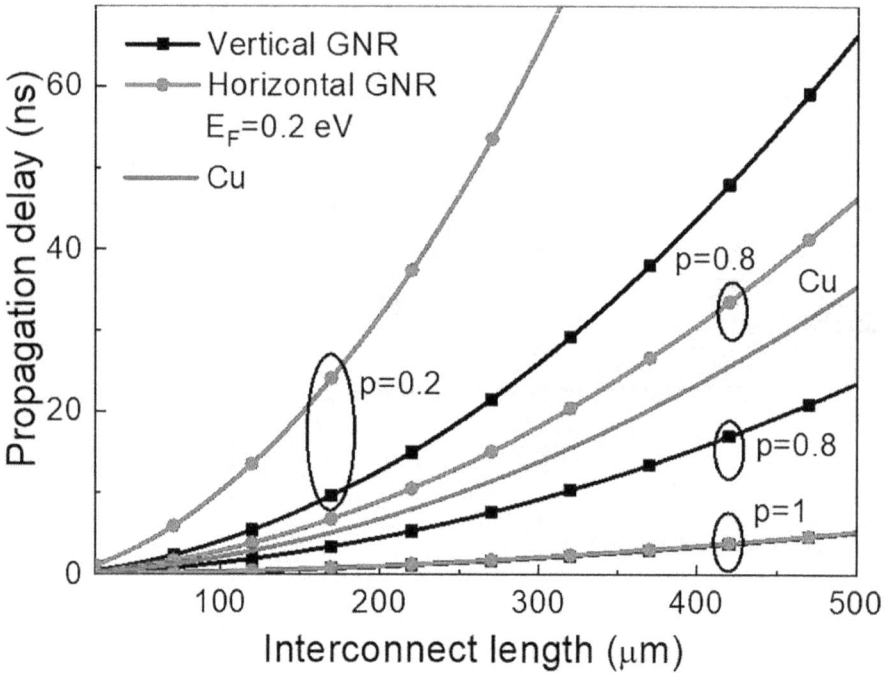

FIGURE 2.19 Delay of the copper, horizontal GNR, and vertical GNR interconnects at 7.5 nm node [46].

and vertical GNRs. In particular, for $p = 0.8$, the vertical GNR interconnect is superior to both copper and horizontal GNR interconnects. To further improve the electrical performance of GNR interconnects, various doping techniques have been developed in the past decade [48–50]. Figure 2.20 shows the propagation delay of pristine and doped vertical GNR interconnects at the 7.5 nm technology node. The physical parameters of the doped GNR interconnects are adopted from [51]. It is evident that appropriate doping techniques could significantly reduce the propagation delay of the vertical GNR interconnects.

As aforementioned, the vertical GNR interconnect is expected to facilitate heat dissipation in ICs. Figure 2.21 shows the unit cell for simulation together with the simulated temperature rise. In the simulation, the temperatures at the top and bottom layers of the unit cell are set as 378 and 318 K, respectively. The geometrical parameters are $W = 7.5$ nm, $T = 15.75$ nm, and $T_{ox} = 14.25$ nm for intermediate level; $W = 11.5$ nm, $T = 26.91$ nm, and $T_{ox} = 17.25$ nm for global level. The thermal conductivities of the inter-metal dielectric and interlayer dielectric are 0.12 and 8 W/m·K, respectively [52]. The equivalent packaging layer is utilized for simplifying the multilayer packaging layer, and its thickness and thermal conductivity are 70 nm and 10^{-5} W/m·K. The power is input into the interconnect at topmost layer. It is evident that horizontal GNR interconnect would hinder the heat dissipation although GNRs

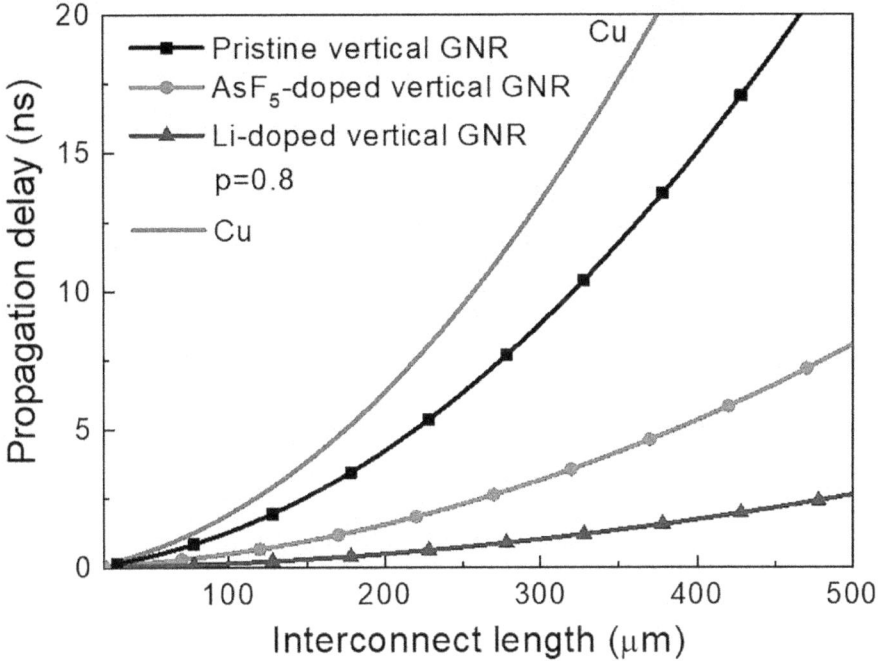

FIGURE 2.20 Delay of the copper, pristine vertical GNR, and doped vertical GNR interconnects at 7.5 nm node [46].

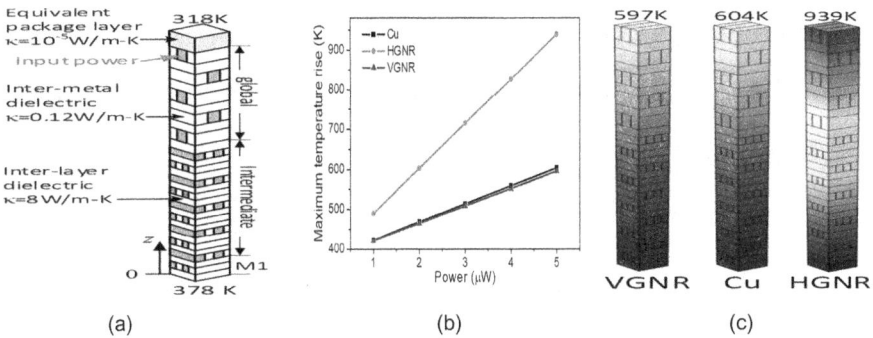

FIGURE 2.21 (a) Unit cell for simulation; (b) maximum temperature rise vs input power; and (c) temperature profile of the unit cell with a power of 5 µW input at the topmost global interconnect [46].

are promising interconnect conductor material due to their high ampacity. The implementation of vertical GNR could avoid such thermal problem, which makes the vertical GNR interconnect suitable for applications in future monolithic 3D ICs [53].

2.6 CONCLUSION

With continuous downscaling of feature dimensions, the interconnect resistivity increases dramatically, thereby resulting in serious performance degradation and reliability problems. Graphene could be a promising candidate as conductor material for interconnect applications. In this chapter, circuit modeling and performance analysis of graphene nanoribbon interconnects have been performed. The number of effective conducting channels of graphene nanoribbon has been extracted numerically. The transmission line circuit model of monolayer graphene nanoribbon interconnect has been developed. To reduce resistive loss, the multilayer graphene nanoribbon interconnect has been explored. Finally, the vertical graphene nanoribbon interconnect scheme has been presented and its performance has been evaluated.

ACKNOWLEDGMENTS

This study was supported by the National Natural Science Foundation of China under Grant 62222401 and the Zhejiang Provincial Natural Science Foundation under Grant LXR22F040001.

REFERENCES

[1] K. Banerjee, S. J. Souri, P. Kapur, and K. C. Saraswat, 3-D ICs: A novel chip design for improving deep-submicrometer interconnect performance and systems-on-chip integration, *Proc. IEEE* vol. 89, no. 5, pp. 602–633, 2001.

[2] M. M. El-Desouki, S. M. Abdelsayed, M. J. Deen, N. K. Nikolova, and Y. M. Haddara, The impact of on-chip interconnections on CMOS RF integrated circuits, *IEEE Trans. Electron Devices* vol. 56, no. 9, pp. 1882–1890, 2009.

[3] D. Josell, S. H. Brongersma, and Z. Tokei, Size-dependent resistivity in nanoscale interconnects, *Ann. Rev. Mater. Res.* vol. 39, pp. 231–254, 2009.

[4] A. A. Vyas, C. Zhou, and C. Y. Yang, On-chip interconnect conductor materials for end-of-roadmap technology nodes, *IEEE Trans. Nanotechnol.* vol. 17, no. 1, pp. 4–10, 2018.

[5] R. Zhang, T. Liu, K. Yang, and L. Milor, CacheEM: For reliability analysis on Cache memory aging due to electromigration, *IEEE Trans. Comput. Aided Des. Integr. Circ. Syst.* vol. 41, no. 9, pp. 3078–3091, 2022.

[6] R. Zhang, K. X. Yang, T. Z. Liu, and L. Milor, Modeling of FinFET SRAM array reliability degradation due to electromigration, *Microelectron. Reliab.* vol. 100, p. 113485, 2019.

[7] W. Y. Yin and W. S. Zhao, *Modeling and Characterization of On-Chip Interconnects, Wiley Encyclopedia of Electrical and Electronics Engineering*, Wiley, New York, 2013.

[8] J. D. Meindl, Beyond Moore's law: The interconnect era, *Comput. Sci. Eng.* vol. 5, no. 1, pp. 20–24, 2003.

[9] K. S. Novoselov, A. K. Geim, S. V. Morozov, D. Jiang, Y. Zhang, S. V. Dubonos, I. V. Grigorieva, and A. A. Firsov, Electric field effect in atomically thin carbon films, *Science* vol. 306, no. 5696, pp. 666–669, 2004.

[10] D. S. L. Abergel, V. Apalkov, J. Berashevich, K. Ziegler, and T. Chakraborty, Properties of graphene: A theoretical perspective, *Adv. Phys.* vol. 59, no. 4, pp. 261–482, 2010.

[11] A. Maffucci and G. Miano, Electrical properties of graphene for interconnect applications, *Appl. Sci.* vol. 4, no. 2, pp. 305–317, 2014.

[12] W.-S. Zhao and W.-Y. Yin, *Carbon-Based Interconnects for RF Nanoelectronics, Wiley Encyclopedia of Electrical and Electronics Engineering*, Wiley, New York, 2012.

[13] Y. Wu, D. B. Farmer, F. Xia, and P. Avouris, Graphene electronics: Materials, devices, and circuits, *Proc. IEEE* vol. 101, no. 7, pp. 1620–1637, 2013.

[14] W.-S. Zhao, K. Fu, D.-W. Wang, M. Li, G. Wang, and W.-Y. Yin, Mini-review: Modeling and performance analysis of nanocarbon interconnects, *Appl. Sci.* vol. 9, p. 2174, 2019.

[15] R. Murali, Y. Yang, K. Brenner, T. Beck, and J. D. Meindl, Breakdown current density of graphene nanoribbons, *Appl. Phys. Lett.* vol. 94, no. 24, p. 243114, 2009.

[16] A. Behnam, A. S. Lyons, M.-H. Bae, E. K. Chow, S. Islam, C. M. Neumann, and E. Pop, Transport in nanoribbon interconnects obtained from graphene grown by chemical vapor deposition, *Nano Lett.* vol. 12, no. 9, pp. 4424–4430, 2012.

[17] J. Jiang, J. Kang, W. Cao, X. Xie, H. Zhang, J. H. Chu, W. Liu, and K. Banerjee, Intercalation doped multilayer-graphene-nanoribbons for next-generation interconnects, *Nano Lett.* vol. 17, no. 3, pp. 1482–1488, 2017.

[18] J. Jiang, J. Kang, J. H. Chu, and K. Banerjee, All-carbon interconnect scheme integrating graphene-wires and carbon-nanotube-vias, in *IEEE International Electron Devices Meeting (IEDM)*, IEEE, San Francisco, CA, 2017.

[19] Y. Zhu, C. W. Tan, S. L. Chua, Y. D. Lim, B. Vaisband, B. K. Tay, E. G. Friedman, and C. S. Tan, Assembly process and electrical properties of top-transferred graphene on carbon nanotubes for carbon-based 3-D interconnects, *IEEE Trans. Comp. Pack. Manuf. Technol.* vol. 10, no. 3, pp. 516–524, 2020.

[20] Y.-F. Liu, W.-S. Zhao, Z. Yong, Y. Fang, and W.-Y. Yin, Electrical modeling of three-dimensional carbon-based heterogeneous interconnects, *IEEE Trans. Nanotechnol.* vol. 13, no. 3, pp. 488–495, 2014.

[21] N. Li, J. Mao, W.-S. Zhao, M. Tang, W. Chen, and W.-Y. Yin, Electrothermal cosimulation of 3-D carbon-based heterogeneous interconnect, *IEEE Trans. Comp. Pack. Manuf. Technol.* vol. 6, no. 4, pp. 518–526, 2016.

[22] G. W. Hanson, *Fundamentals of Nanoelectronics*, Pearson, India, 2019.

[23] A. C. Neto, F. Guinea, N. M. Peres, K. S. Novoselov, and A. K. Geim, The electronic properties of graphene, *Rev. Modern Phys.* vol. 81, no. 1, p. 109, 2009.

[24] A. Maffucci and G. Miano, Number of conducting channels for armchair and zig-zag graphene nanoribbon interconnects, *IEEE Trans. Nanotechnol.* vol. 12, no. 5, pp. 817–823, 2013.

[25] C. Berger, Z. Song, X. Li, X. Wu, N. Brown, C. Naud, D. Mayou, T. Li, J. Hass, A. N. Marchenkov, and E. H. Conrad, Electronic confinement and coherence in patterned epitaxial graphene, *Science* vol. 312, no. 5777, pp. 1191–1196, 2006.

[26] A. Naeemi and J. D. Meindl, Compact physics-based circuit models for graphene nanoribbon interconnects, *IEEE Trans. Electron Devices* vol. 56, no. 9, pp. 1822–1833, 2009.

[27] S. Rakheja, V. Kumar, and A. Naeemi, Evaluation of the potential performance of graphene nanoribbons as on-chip interconnects, *Proc. IEEE* vol. 101, no. 7, pp. 1740–1765, 2013.

[28] W.-S. Zhao and W.-Y. Yin, Comparative study on multilayer graphene nanoribbon (MLGNR) interconnects, *IEEE Trans. Electromag. Compat.* vol. 56, no. 3, pp. 638–645, 2014.

[29] A. Naeemi and J. D. Meindl, Conductance modeling for graphene nanoribbon (GNR) interconnects, *IEEE Electron Device Lett.* vol. 28, no. 5, pp. 428–431, 2007.

[30] C. Xu, H. Li, and K. Banerjee, Modeling, analysis, and design of graphene nano-ribbon interconnects, *IEEE Trans. Electron Devices* vol. 56, no. 8, pp. 1567–1578, 2009.

[31] F. Stellari and A. L. Lacaita, New formulas of interconnect capacitances based on results of conformal mapping method, *IEEE Trans. Electron Devices* vol. 47, no. 1, pp. 222–231, 2000.

[32] K. I. Bolotin, K. J. Sikes, Z. Jiang, M. Klima, G. Fudenberg, J. Hone, P. Kim, and H. L. Stormer, Ultrahigh electron mobility in suspended graphene, *Solid State Commun.* vol. 146, no. 9, pp. 351–355, 2008.

[33] X. Wang, Y. Ouyang, L. Jiao, H. Wang, L. Xie, J. Wu, J. Guo, and H. Dai, Graphene nanoribbons with smooth edges behave as quantum wires, *Nat. Nanotechnol.* vol. 6, no. 9, pp. 563–567, 2011.

[34] V. R. Kumar, M. K. Majumder, N. R. Kukkam, and B. K. Kaushik, Time and frequency domain analysis of MLGNR interconnects, *IEEE Trans. Nanotechnol.* vol. 14, no. 3, pp. 484–492, 2015.

[35] M. Sanaeepur, Dielectric surface roughness scattering limited performance of MLGNR interconnects, *IEEE Trans. Electromag. Compat.* vol. 61, no. 2, pp. 532–537, 2019.

[36] Y. Agrawal, M. G. Kumar, and R. Chandel, A novel unified model for copper and MLGNR interconnects using voltage- and current-mode signaling schemes, *IEEE Trans. Electromag. Compat.* vol. 59, no. 1, pp. 217–227, 2017.

[37] V. Kumar, S. Rakheja, and A. Naeemi, Performance and energy-per-bit modeling of multilayer graphene nanoribbon conductors, *IEEE Trans. Electron Devices* vol. 59, no. 10, pp. 2753–2761, 2012.

[38] W.-S. Zhao, D.-W. Wang, G. Wang, and W.-Y. Yin, Electrical modeling of on-chip Cu-graphene heterogeneous interconnects, *IEEE Electron Device Lett.* vol. 36, no. 1, pp. 74–76, 2015.

[39] M. S. Sarto and A. Tamburrano, Comparative analysis of TL models for multilayer graphene nanoribbon and multiwall carbon nanotube interconnects, in *IEEE International Symposium on Electromagnetic Compatibility*, IEEE, Lauderdale, FL, 2010.

[40] J.-P. Cui, W.-S. Zhao, W.-Y. Yin, and J. Hu, Signal transmission analysis of multilayer graphene nano-ribbon (MLGNR) interconnects, *IEEE Trans. Electromag. Compat.* vol. 54, no. 1, pp. 126–132, 2011.

[41] A. G. D'Aloia, W.-S. Zhao, G. Wang, and W.-Y. Yin, Near-field radiated from carbon nanotube and graphene-based nanointerconnects, *IEEE Trans. Electromag. Compat.* vol. 59, no. 2, pp. 646–653, 2017.

[42] Z.-H. Cheng, W.-S. Zhao, D.-W. Wang, J. Wang, L. Dong, G. Wang, and W.-Y. Yin, Analysis of Cu-graphene interconnects, *IEEE Access* vol. 6, pp. 53499–53508, 2018.

[43] Y. I. Ismail and E. G. Friedman, Effects of inductance on the propagation delay and repeater insertion in VLSI circuits, *IEEE Trans. Very Large Scale Integr. (VLSI) Syst.* vol. 8, no. 2, pp. 195–206, 2000.

[44] E. Pop, V. Varshney, and A. K. Roy, Thermal properties of graphene: Fundamentals and applications, *MRS Bull.* vol. 37, no. 12, pp. 1273–1281, 2012.

[45] M. Nihei, A. Kawabata, T. Murakami, M. Sato, and N. Yokoyama, Improved thermal conductivity by vertical graphene contact formation for thermal TSV, in *IEEE International Electron Devices Meeting (IEDM)*, San Francisco, CA, 2012.

[46] W.-S. Zhao, Z.-H. Cheng, J. Wang, K. Fu, D.-W. Wang, P. Zhao, G. Wang, and L. Dong, Vertical graphene nanoribbon interconnects at the end of the roadmap, *IEEE Trans. Electron Devic.* vol. 65, no. 6, pp. 2632–2637, 2018.

[47] ITRS, *International Technology Roadmap for Semiconductors (ITRS)*, https://itrs.net/, 2018.

[48] H. Terrones, R. Lv, M. Terrones, and M. S. Dresselhaus, The role of defects and doping in 2D graphene sheets and 1D nanoribbons, *Rep. Prog. Phys.* vol. 75, no. 6, p. 062501, 2012.

[49] E. Carbonell-Sanroma, J. Hieulle, M. Vilas-Varela, P. Brandimarte, M. Iraola, A. Barragan, J. Li, M. Abadia, M. Corso, D. Sanchez-Portal, and D. Pena, Doping of graphene nanoribbons via functional group edge modification, *ACS Nano* vol. 11, no. 7, pp. 7355–7361, 2017.

[50] F. Joucken, Y. Tison, P. Le Fevre, A. Tejeda, A. Taleb-Ibrahimi, E. Conrad, V. Repain, C. Chacon, A. Bellec, Y. Girard, and S. Rousset, Charge transfer and electronic doping in nitrogen-doped graphene, *Sci. Rep.* vol. 5, no. 1, pp. 1–10, 2015.

[51] A. K. Nishad and R. Sharma, Lithium-intercalated graphene interconnects: Prospects for on-chip applications, *IEEE J. Electron Devices Soc.* vol. 4, no. 6, pp. 485–489, 2016.

[52] S. Im, N. Srivastava, K. Banerjee, and K. E. Goodson, Scaling analysis of multilevel interconnect temperatures for high performance ICs, *IEEE Trans. Electron Devices* vol. 52, no. 12, pp. 2710–2719, 2005.

[53] M. M. Shulaker, G. Hills, R. S. Park, R. T. Howe, K. Saraswat, H.-S. P. Wong, and S. Mitra, Three-dimensional integration of nanotechnologies for computing and data storage on a single chip, *Nature* vol. 547, pp. 74–78, 2017.

3 Introduction to Nanoscale Interconnect Materials

Laxman Raju Thoutam and Shantikumar Nair

3.1 INTRODUCTION

Integrated circuit (IC) technology plays a pivotal role in the design and development of modern nanoscale electronic devices. The continuous decrease of transistor size over the years to keep up with Moore's law prediction seems to reach its physical dimensional limits. The current sub-nanometre technology demands critical control of material properties at various levels of IC making process; sophisticated lithographical tools and optimized fabrication protocols yield reliable nanoscale devices with optimum performance for functional applications. Silicon-based technologies, the frontier of all modern electronic devices, also seem to face a ubiquitous challenge of physical down-scaling of transistor size, energy efficiency, power handling and memory limitations [1,2]. The current research is focussed on discovering, designing and developing alternate material systems that can compete or outperform conventional silicon-based materials at nanoscale to offer high integrated circuit package density, low power consumption and fast speeds. Please note that the emerging material systems should be compatible with existing fab technologies for process optimization and allow ease of integration to control the production cost. Two-dimensional (2D) materials including graphene [3], transition metal dichalcogenides [4,5], MXenes [6,7] and other atomic flat mono-layers like phosphorous [8,9] and silicene [10] are considered as some of the potential candidates to replace silicon as semiconductor channel material at the nanoscale. The search for new semiconductor channel material also mandates exploration of new and novel interconnect materials that link different electronic nodes in an integrated circuit. It needs to be emphasized that decreasing transistor size and increasing the number of transistors for a given physical space invariably increase the complexity and number of interconnects to maintain desired functionality. The present interconnect technology seems to face a difficult challenge of critical dimensional control and understanding the nanoscale size effects on the performance of the interconnect structures [11]. For instance, the decrease in line width of the interconnect can lead to enhanced surface scattering and grain boundary scattering that increase the interconnect's resistivity, which can lead to thermal losses and decrease in power efficiency [12,13]. For example, the most commonly used interconnect material copper's resistivity increases by an order of

DOI: 10.1201/9781003331650-3

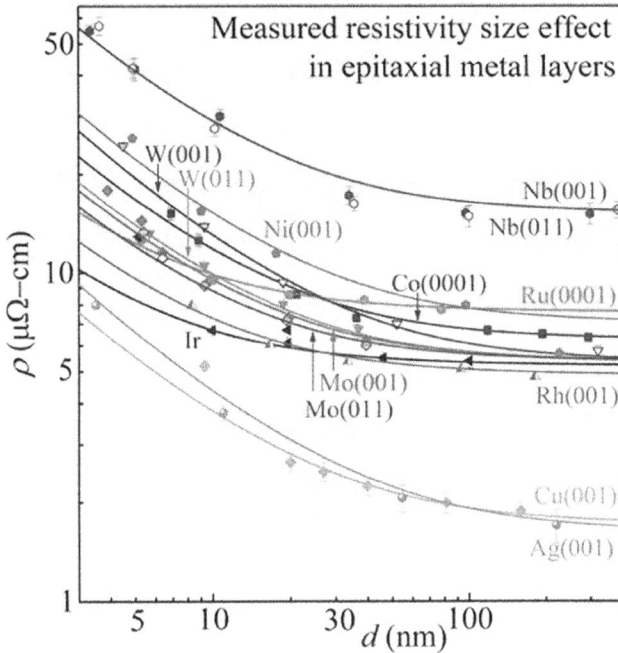

FIGURE 3.1 The scaling of room-temperature experimental resistivity vs thickness of the commonly used elemental metals as thin film layers. Permission from D. Gall et al., *J. Appl. Phys.* 127, 050901, 2020.

magnitude when its dimensions are reduced from bulk to a 10-nm-wide line [14]. The increase in resistivity will also lead to high RC values in interconnect systems which increases the overall interconnect delays of the integrated circuit. A comparison of room-temperature experimental measured resistivity vs thickness of the epitaxial thin films of elemental metals is presented in Figure 3.1. It is clear that copper and silver have the smallest resistivity values at the bulk (see Figure 3.1). It is to be noted that silver suffers from agglomeration and corrosion, and often requires an encapsulation layer to retain its intrinsic properties, which poses serious reliability issues in IC design and testing [15].

It is obvious that copper resistivity increases by an order of magnitude with reduced dimensions and this increase in resistivity with reduced dimensions is seen in all other metals including ruthenium, iridium, molybdenum, cobalt, nickel and niobium [16]. However, the amount of change in resistivity differs for different materials and their crystal orientation, hinting about the possibility of material engineering as a go-to technique to mitigate the common interconnect issues that arise at reduced nanoscale dimensions. This chapter highlights the key limiting factors of current nanoscale interconnect materials. This chapter outlines the effect of surface scattering, grain boundary scattering and the interface quality on size effects, i.e., increase in resistivity at reduced dimensions for widely used copper interconnect. This chapter briefly discusses various emerging nanoscale interconnect materials systems including elemental conductive metals, carbon-based materials, transition metal dichalcogenides and MXenes. This chapter showcases the prospects of different nanoscale emerging interconnect systems to be able to integrate into future sub-nanometre technologies.

3.2 ISSUES IN NANOSCALE INTERCONNECTS

The wide use of copper as interconnect material in current complementary metal-oxide semiconductor (CMOS) technology is due to its low resistivity [16], high mechanical strength to withstand the stress under chip-package interaction loads and multi-level interconnections [17–19], high current-carrying capacity to mitigate Joule heating [20], chemical stability and high melting point to yield high electromigration resistance [21], and it is easily integrable and process-friendly for most fabrication protocols [22,23]. However, as presented in Figure 3.1, copper's resistivity increases as we move towards sub-nanometre technological nodes. The decrease in physical dimensions of copper at nanoscale increases its surface-to-volume ratio, enabling higher scattering affects at surface and grain boundaries raising a serious question about its feasibility and viability at lower technological nodes. The fundamental understanding of different scattering effects and their physical origins/mechanisms will help to tailor the future emerging material systems to adapt and suit nanoscale interconnect technology. The surface and grain boundary scattering play a major role and dictate the resistivity values of interconnect materials at nanoscale [16]. At nanoscale, the mean free path of the electrons in a material becomes comparable to the interconnect's physical dimensions and any surface effects would profoundly impact the interconnect's resistivity [24]. The reduced dimensionality also puts stringent restrictions of fabrication protocols to realize high-aspect-ratio structures with optimum properties.

3.2.1 Surface Scattering

The electron surface scattering in metallic thin film interconnects can be analysed by Fuchs-Sondheimer (FS) analytical model, which uses Boltzmann transport equations [25,26]. The model incorporates surface scattering as a boundary condition and introduces a phenomenological scattering specular parameter 'p'. The value of p dictates the type of scattering and it lies between 0 (diffusive scattering) and 1 (specular scattering). In diffusive scattering, the surface scattering randomizes the electron's momentum resulting in increase of resistivity, see Figure 3.2a. On the other hand, specular scattering refers to electron reflected by the surface layer with conservation of parallel momentum, yielding no change in the resistivity as shown in Figure 3.2b [24,27].

Specular reflection at the surface boundaries requires the growth of atomic flat smooth layers, which requires stringent and optimum growth protocols via atomic layer deposition and molecular beam epitaxial methods. Kuan et al. proposed a simplified FS model to calculate interconnect resistivity taking experimental interconnect parameter thickness 't' into consideration and is given as [25,26,28]

(a) Diffuse Scattering (p = 0)

Increase in resistivity

(b) Specular Scattering (p = 1)

No change in resistivity

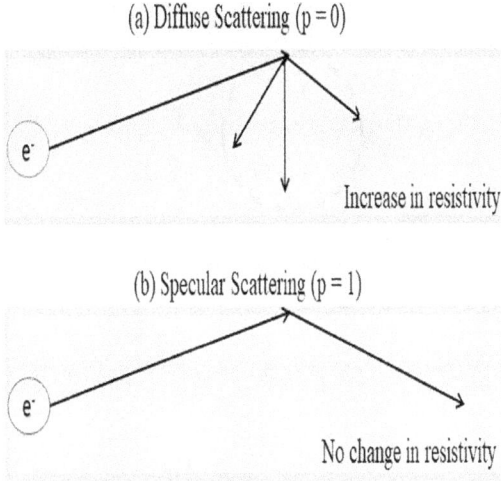

FIGURE 3.2 Schematic illustration of (a) diffuse scattering and (b) specular scattering. The grey lines denote the surface boundaries and the black arrowed lines correspond to electron path. Adapted from [27].

$$\frac{\rho_{film}}{\rho_{bulk}} \approx 1 + \frac{0.375l(1-p)}{t} \tag{3.1}$$

where ρ_{film} and ρ_{bulk} represent the resistivities of the interconnect material in epitaxial thin film and bulk, respectively; l is the mean free path; and p is the specular parameter. Copper's calculated resistivity as a function of film thickness 't' (as per equation 3.1) clearly shows (see Figure 3.3) the sharp increase of resistivity at smaller thicknesses ($t < 45$ nm) in diffuse scattering regime ($p = 0$) as opposed to no change in resistivity for specular scattering ($p = 1$).

The presence of impurities or adatoms, and any surface roughness created during the fabrication process can also act as potential scattering centres on the surface of the materials and will lead to considerable change in the resistivity. Techniques like encapsulation can help to minimize the surface adatom scattering, albeit at the expense of complex Fermi surfaces of the encapsulation layers should be critically addressed for the choice of overlayer material [29]. For example, a 10% rise in sheet resistance is observed when 5 nm tantalum is overlaid on copper. On the other hand, a decrease in sheet resistance for copper is observed when it is encapsulated with aluminium metal [29]. Thus, the choice of encapsulated material can be a deciding factor to overcome the size effects of resistivity in copper interconnect technology. However, the encapsulated materials like tantalum and aluminium are sensitive to ambient conditions and are easily oxidized, raising a serious question on the reliability of the interconnects. Additionally, any overlayer material introduces additional steps in the interconnect fabrication process and thereby increases the production cost. The search for alternate material systems with larger mean free paths and averse to size effects would be beneficial for future nanoscale interconnect technologies.

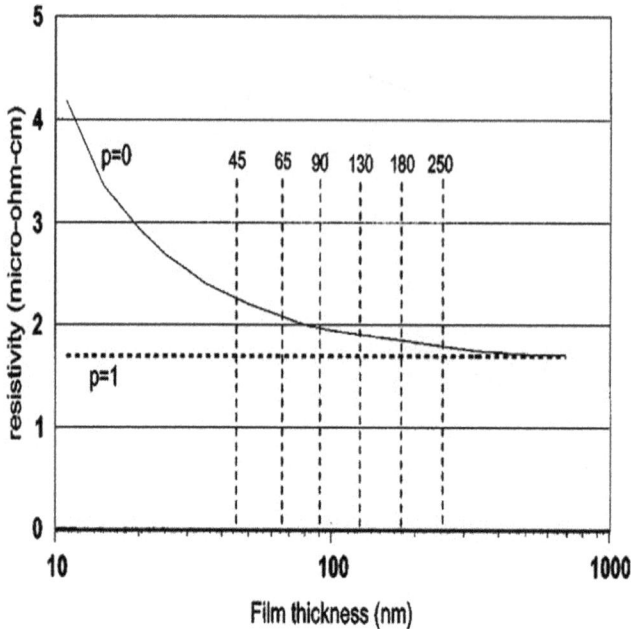

FIGURE 3.3   The calculated resistivity of copper vs thickness of the films as per the Kaun model presented in equation (1.1). The resistivity increases with decreasing thickness for diffuse scattering ($p = 0$) and stays constant for specular scattering ($p = 1$) at all thicknesses. The vertical dotted lines represent the commercial technological nodes of fabrication. Permission from S. M. Rossnagel et al., *J. Vac. Sci. Technol. B.* 22, 240, 2004.

3.2.2   INTERFACE TYPE AND QUALITY

The continued decrease of interconnect line widths to facilitate higher circuit package density will give rise to thermal management issues. Thermal boundary resistance (TBR), defined as the resistance between the interconnect and the surrounding environment, is a significant factor that contributes to overall thermal resistance of the interconnect [30]. The thermal resistance leads to rise in local temperature of the interconnect and can lead to electromigration and stress migration effects [31]. The nanoscale interconnect's high surface-to-volume ratio will facilitate the formation of large-scale interfaces between the constituent materials. The thermal resistance of different interfaces including metal/metal, metal/dielectric, dielectric/dielectric has to be individually analysed to mitigate thermal losses [32]. A typical interconnect structure shown in the inset of Figure 3.4 highlights the presence of different interfacial layers. For example, in a conventional copper damascene interconnect structure, a barrier layer of TaN is used to prevent copper diffusion into surrounding interlayers and also acts as a protection layer to prevent surface oxidation of copper. The poor adhesion between copper and TaN requires the use of Ta liner layer [33]. The high current flowing through the nanoscale interconnects increases the local temperature that has to be dissipated through the several interfaces before reaching the low-k dielectric interlayer [32].

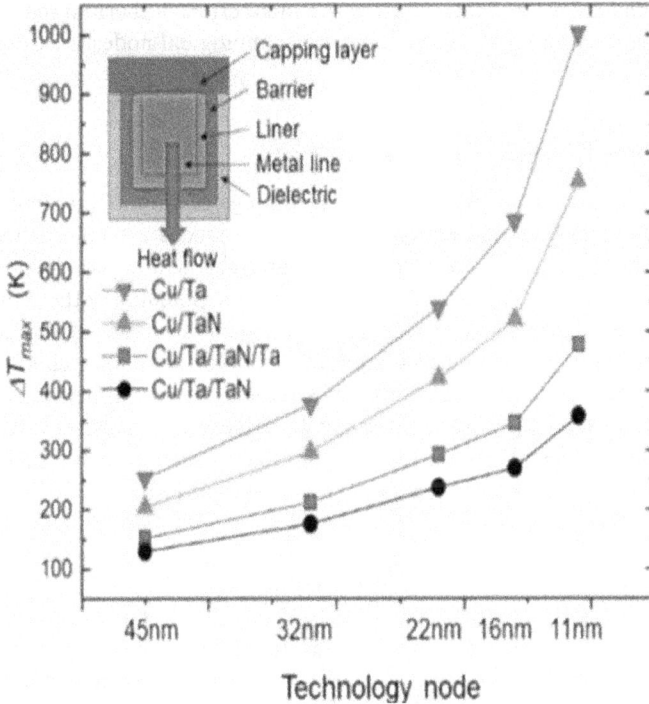

FIGURE 3.4 The variation in the temperature of copper/liner/barrier/SiO$_2$ structure with different liner/barrier layers at representative technological nodes. The insets show the cross-sectional view of metal/liner/barrier/dielectric copper damascene interconnect structure. Permission from T. Zhan et al., *ACS Appl. Mater. Interfaces* 12, 22347–22356, 2020.

The physical origin of the TBR between different interfaces including metal/dielectric and dielectric/dielectric is due to phonon density of states (DOS) mismatch at either side of the interface [34]. The higher the mismatch of DOS, the larger the TBR. The metal/metal interface has a high chance of localized DOS that acts as potential scattering centres and leads to diffusive scattering leading to increased resistivity. On the other hand, a dielectric layer typically has negligible DOS at the Fermi level and hosts small numbers of scattering centres at the metal/dielectric interface favouring specular scattering [16]. Zhan et al. have studied the effect of different barrier layer on copper/barrier/SiO$_2$ structure to estimate the rise in temperature for a global-tier interconnect using finite element method (FEM) analysis. It is proposed that the interconnect temperature rises with down-scaling of technological nodes (see Figure 3.4), and a significant rise of temperature greater than 300 K at 11 nm interconnect technology is observed for the widely used Cu/Ta/TaN/SiO$_2$ structure [32]. The interface quality of the interconnect material is also dependent on surface roughness of the interconnect materials, interdiffusion at the interface layer, bonding strength and interface chemistry, which typically are process-dependent [32]. These signify the search for alternate materials systems with clean and defect-free interfaces to facilitate efficient thermal management and easy process integration at sub-nanometre technological nodes for future interconnect systems.

3.2.3 GRAIN BOUNDARY SCATTERING

The grain boundary scattering is due to the change of atomic structure of the grain boundary compared to its interior. The change induces a local change in atomic potential from its adjacent areas across the interface, leading to electron scattering at the grain boundary. The deviation of local atomic potential at the grain boundary interface is dependent on local atomic arrangement, structure of the grain boundary and bonding [35]. The narrow interconnect lines at small technological nodes hosts grain boundaries that are comparable to their physical dimensions and thereby see increased scattering to result in a net increase in the resistivity. The interconnect's resistivity dependence on the grain boundary scattering can be estimated by the Mayadas and Shatzkes (MS) model and is given as [36]

$$\rho = \rho_{bulk}l\left(\frac{3R}{2D(1-R)}\right) \tag{3.2}$$

where ρ and ρ_{bulk} are the resistivity in thin film and bulk material, respectively; l is the mean free path, D is the grain size; and R is a phenomenological reflection coefficient which measures the probability of an electron being backscattered by a grain boundary. The grain boundary scattering can be reduced by increasing the grain size as per equation (3.2). For instance, if the grain size exceeds 200 nm (and is larger than mean free path), the electron scattering at grain boundary is considerably decreased (<10%) for a typical metal with $l = 30$ nm and $R = 30\%$ [16]. Sun et al. have experimentally studied the grain size effects on copper and observed a nominal change in resistivity (<10%) for grain sizes of $D > 200$ nm; see Figure 3.5 [37]. The solid line is the prediction of MS model (equation 3.2) with $R = 0.47$ to the experimentally measured resistivity values at different temperatures. The experimental room-temperature change in resistivities of copper (solid symbols in Figure 3.5) is always greater than low-temperature values (open symbols in Figure 3.5), and this dependence is not accounted by the MS model [37].

The phenomenological reflection coefficient 'R' is an intrinsic material property and is a strong function of boundary geometry. For example, a twin boundary has $R = 0.016$ [38] compared to random boundaries with $R = 0.65-0.75$ [39,40]. The control over growth kinetics and parameters during the fabrication of nanoscale interconnects should be optimized to result in the formation of a symmetrical low-energy boundary vs a randomized high-energy boundary that induces high potential changes at the grain boundary favouring large scattering, leading to substantial increase in the resistivity [35]. Alternatively, the potential fluctuations at the grain boundary sites can be compensated by introducing controlled amounts of external dopants into the host interconnect material [41]. Caser et al. have studied the effect of dopants on grain boundary resistivity using self-consistent nonequilibrium Green's function within density functional theory (NEGF-DFT) framework and nonequilibrium

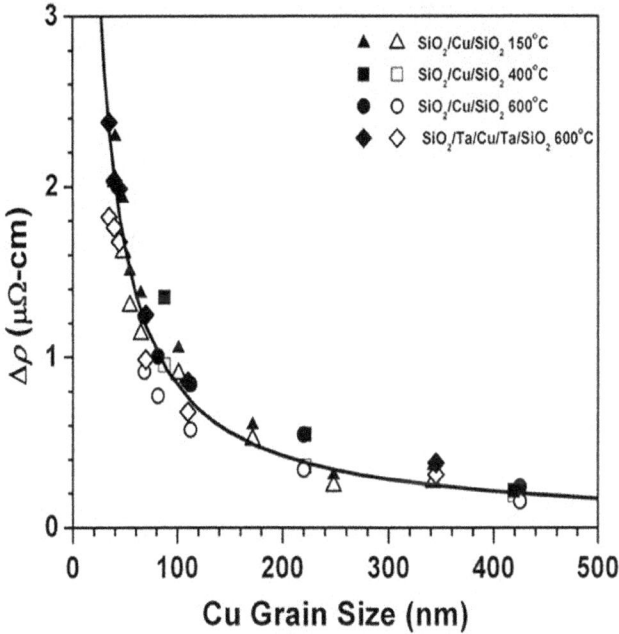

FIGURE 3.5 The net change in increase of resistivity (over its bulk value) vs grain size of copper in $SiO_2/Cu/SiO_2$ and $SiO_2/Ta/Cu/SiO_2$ structures at different constant temperature (closed symbols) and open symbols denote the data points at $T = 4.2$ K. Please note that the mean free path of copper at room temperature is ~39 nm. Open access from T. Sun, "Classical size effect in copper thin films: Impact of surface and grain boundary scattering on resistivity," Electronic Theses and Dissertations, 3927, 2009.

coherent potential approximation (NECPA) and proposed silver, zinc, magnesium, palladium, aluminium and indium can potentially serve as good charge compensating dopants in copper [41]. However, it should be noted that incorporation of external dopants into current copper-based interconnect technology requires modification of existing process protocols. On the other hand, the external dopant atoms if not targeted properly to be at the grain boundary site will lead to additional complexities like modification of density of sites and structural imperfections to alter the interconnect properties.

It seems plausible in all scenarios that research should be focussed on to explore alternate material systems that can potentially replace copper and its associated barrier layers to mitigate size effects at nanoscale. Research on the new emerging nanoscale interconnect material systems should also consider the abundancy and availability of the material that invariably controls the production cost. However, the newer materials systems, either elemental or combinatorial, require new processes that increase the integration complexity [42]. The maximum current carrying capacity of an interconnect is primarily restricted by its fundamental material properties and physical cross-sectional dimensions. It should be noted that the current carrying capacity of the nanoscale interconnect materials is also dependent on barrier and dielectric layer properties. Additionally, incorporation of quantum transport models in existing surface and grain boundary scattering models (equations 3.1 and 3.2, respectively) would yield a wealth of information about the performance of future interconnect materials [16]. The use of latest computational techniques like artificial intelligence and machine learning computational techniques can be best put in practice to select the ideal metal/barrier layer for optimum interconnect performance. Thus, a comprehensive understanding of emerging interconnect materials and their interaction with different barrier/dielectric layer interfaces is much needed to foresee the feasibility of integration with current and future semiconductor technologies.

3.3   EMERGING NANOSCALE INTERCONNECT MATERIALS

The current copper-based interconnects require a barrier/liner layer to prevent copper diffusion into surrounding layers, affecting the reliability of the IC. The down-scaling of copper mandates the parallel down-scaling of barrier/liner layers to achieve high circuit package density. However, growth of few nanometre (<2 nm) barrier/liner layer is not trivial and achieving high aspect ratio for the interconnect at sub-nanometre level puts a hard limit on the usage of copper [43]. Innovations and advancements in materials science and fabrication technologies can aid in synthesizing high-quality material with clean interfaces in different architectures to be compatible with future vertical three-dimensional configurations to achieve high circuit package density. The search for alternate material systems with superior nanoscale thermal, chemical, mechanical and electronic properties is required for future nanoscale interconnect technologies.

3.3.1   ELEMENTAL METALS

The elemental metal's resistivity, mean free path and current carrying capacity play a major role in the design and development of future interconnect technologies. All elemental metals show size effect (see Section 3.2.1 and Figure 3.1) with decrease in physical length scales. The choice of the copper replacement is predominantly governed by its ability to mitigate size effects on its intrinsic electronic properties at sub-nanometre technologies [44]. It is to be noted that although copper has the second lowest room-temperature resistivity, its resistivity value reaches comparable levels similar to other metals at reduced dimensions (see Figure 3.1). The increase at low dimensions is a function of its bulk resistivity and mean free path, which can be used as a figure of merit to select copper-alternate materials. The product of room-temperature bulk resistivity and mean free path of various conductive elements is presented in Figure 3.6 [45], and the corresponding melting point of each individual elements is listed inside the histogram [46]. The metallic elements aluminium, rhodium, iridium, molybdenum, nickel, ruthenium and osmium lie below copper's figure of merit (horizontal dashed line in Figure 3.6), suggesting their potential use as future interconnects. However, other physical parameters like melting point of the materials also play a role in selection of alternate materials. A high melting point usually aids in less electromigration, which is a diffusion-based process [47]. For instance, copper's

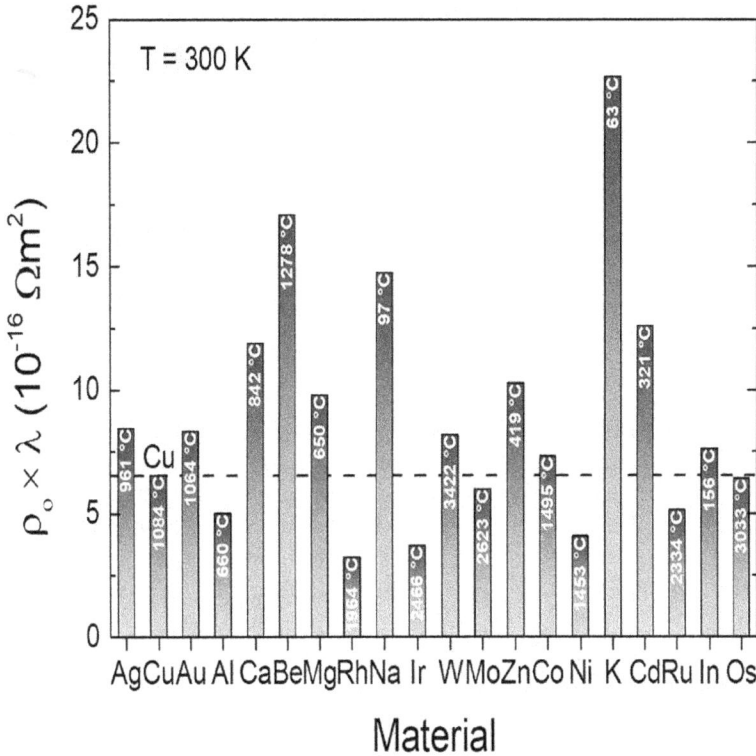

FIGURE 3.6 The calculated product of room-temperature bulk resistivity (ρ_o) and mean free path (λ) of various conductive metals (data is taken from Ref [45]). The melting point of each metal is listed inside the histogram (data is taken from Ref. [46]). The horizontal dashed line is a reference line that represents the widely used copper interconnect's $\rho_o \times \lambda$ data ($6.56 \times 10^{-16} \mu m^2$).

high melting point (1,084°C) over aluminium (660°C) favours less electromigration and improves the reliability of copper interconnects [47].

The elemental metal rhodium has lowest figure of merit and its melting point is 1.8 times higher than copper and should be a good candidate to replace copper interconnect. However, it should be noted that the figure of merit values presented in Figure 3.6 are calculated using first-principles calculation based on classical transport mechanisms and do not account for quantum effects and size effects that occur at nanoscale [48]. Jog et al. recently studied experimental size effects on rhodium and observed a similar increase in resistivity trend with reduced dimensions, as that of copper. However, the resistivity scaling in rhodium is smaller compared to copper and is attributed to smaller mean free path of rhodium, suggesting its possible use as future interconnect [48]. The recent first-principles calculations on rhodium and iridium reveal higher activation energies for diffusion due to their much higher melting points and offer improved reliability by minimizing electromigration reliability issues [49]. Iridium's metal electronics properties strongly depend on the growth method and conditions. For instance, increasing the substrate temperature from 700°C to 1,000°C results in doubling of room-temperature resistivity from 9.78 to 18.7 $\mu\Omega$ cm for a nominal 10-nm-thick iridium epitaxial layer grown on MgO (001) substrate [50]. The growth-dependent issues of iridium must be critically addressed since the interconnect fabrication is a subset of IC fabrication and any process variation would adversely affect iridium's interconnect properties. The molybdenum metal has high bulk resistivity when compared to copper. However, at nanoscale, 8-nm-thin molybdenum's resistivity is lower than TaN/Cu/TaN stack, proving it is less affected by size effects relative to copper and has a huge potential to be used as barrierless interconnect [51]. Molybdenum's extraordinarily high melting point which is two times higher than that of copper should minimize electromigration effects and has full potential to work as reliable nanoscale interconnect for future technologies.

Ruthenium and cobalt are two well-studied metals as nanoscale interconnects due to their reduced resistivity size effects [45], optimized and well-established growth protocols [52–54], high melting points resulting in low electromigration effects and ease of process integration [55]. Choi et al. have performed systematic analysis on the interconnectivity resistivity of ruthenium and cobalt as a function of grain size, film thickness and wire width, and compared them with copper under similar conditions (see Figure 3.7) [56]. The phenomenological scattering parameter 'p' is kept constant at 0.52 and the reflection coefficient 'R' is varied from 0.2 to 0.42 in equations (3.1) and (3.2) respectively. The resistivity size effects are minimal in ruthenium and cobalt when compared to copper at reduced dimensions (<15 nm), as shown in Figure 3.7a and b. However, the size effects in ruthenium and cobalt increase with increase in 'R' value, suggesting the dominant role of grain boundary scattering. For a fixed value of $p = 0.52$ and $R = 0.43$, the absolute magnitude of size effects in ruthenium and cobalt is less compared to copper; and the size effect contribution is decreased by decreasing the grain size compared to wire width for all thicknesses (5–50 nm), as shown in Figure 3.7c and d [56]. The experimental resistivity scaling of ruthenium at reduced dimensions (<15 nm) shows reduced size effects, when compared to copper interconnect in good agreement with the simulations [55,57]. The interfacial bonding of ruthenium with various barrier/liner layers, viz., Ti, Ta, TiN and TaN, was explored and TiN and TaN layers show reduced TBR against standard Cu/Ta/TaN/ SiO$_2$ structure, envisaging better thermal management for them to be used as future nanoscale interconnects [30,43].

3.3.2 CARBON-BASED INTERCONNECTS

The discovery of carbon nanotubes and graphene with superior electronic and mechanical properties has revolutionized the field of nanoscale electronics [58–61]. The International Technology Roadmap for Devices and Systems (ITRS) 2022 report says that interconnect resistance is increasing exponentially at sub-nanometre length scales and currently used copper cannot sustain for future technologies, and put forth the immediate need for copper-alternate material systems [42]. The shrinkage of interconnect sizes has resulted in high current density at nanoscale dimensions and ITRS predicts that it will exceed the break-down limit of copper soon [62]. There is an urgent need for nanoscale materials with high current-carrying capacity. The high

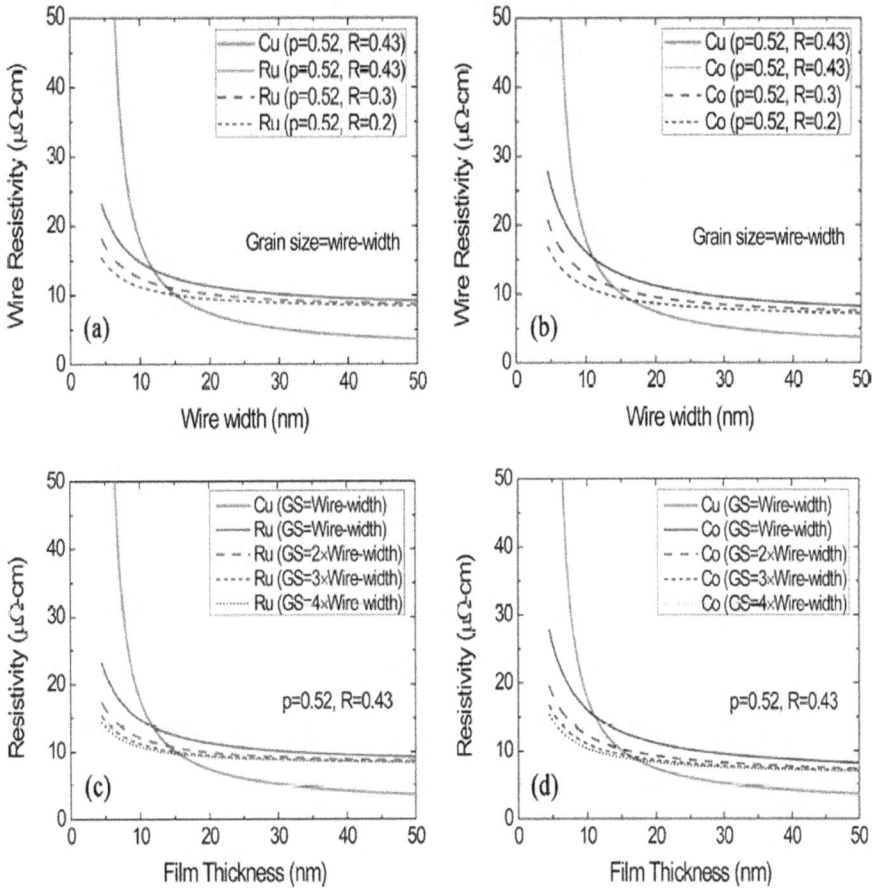

FIGURE 3.7 The calculated resistivity of polycrystalline wires of copper, ruthenium and cobalt as a function of wire width (a, b) and thickness (c, d). The phenomenological parameter p is fixed at 0.52 and R is varied in calculating resistivity using FS model (equation 1.1) and MS model (equation 3.2). Open access CC BY-NC 4.0, D. Choi et al., *Korean J. Met. Mater.* 56, 605–610, 2018.

current-carrying capacity termed as ampacity is a material property and typically depends on its physical bonding structure [63].

The nanoscale carbon varieties including carbon nanotubes (CNT) and graphene are proven to be good electronic conductors and even achieved room-temperature ballistic transport [64,65]. Their excellent nanoscale mechanical and thermal properties aid to achieve high ampacity values needed for future interconnects; see Figure 3.8 [58–61]. The ampacity and conductivity of the interconnects show opposite trends (see Figure 3.8) since ampacity requires a strongly bounded material whereas conductivity requires free electrons and thus weakly bonded system [63]. It is apparent that copper is the most conductive element of all the listed materials in Figure 3.8, but its current carrying capacity is limited. On the other hand, carbon-based interconnects show high current-carrying capacity albeit low conductivity when compared

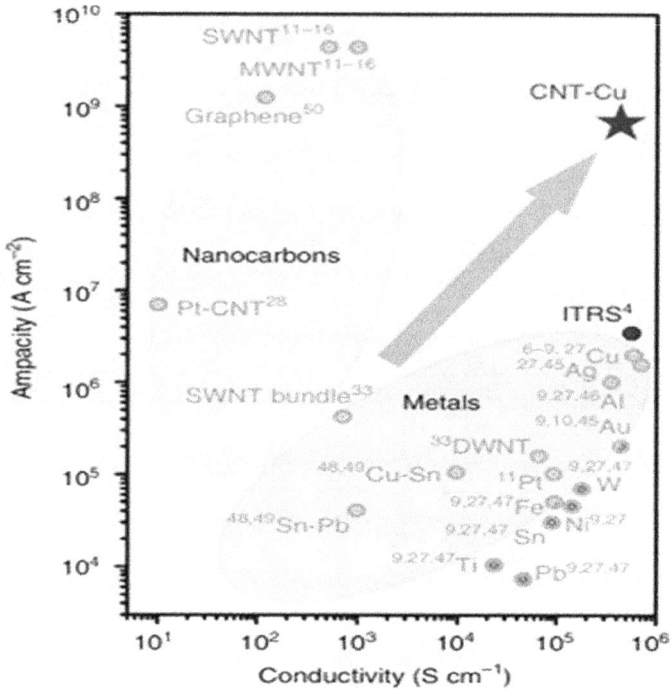

FIGURE 3.8 The ampacity vs conductivity of the various interconnect material families (elemental metals, carbon-based, alloys and composites). Please note that the individual references of each material system listed in the figure can be found at [63]. Open access BY-NC-SA/3.0 C. Subramaniam et al., *Nat. Commun.* 4, 2202, 2013.

to copper. The high current-carrying capacity ($>10^9$ vs 10^6 A/cm^2 for copper), large room-temperature mean free path (65 µm vs 39 nm for copper) and high thermal conductivity of carbon nanotubes (3,000 vs 400 W/m K for copper) would be a useful property at nanoscale, i.e., extremely small CNT interconnect line widths can handle higher current densities with efficient thermal management plan for sub-nanometre interconnect technological nodes [64–71].

However, CNTs can be synthesized/fabricated in many varieties, viz., single walled vs multi-walled, zig-zag vs arm-chair vs chiral and dense CNTs (for more information on synthesis and properties of individual CNTs, please see references [58,72]). The nanometre-sized dimensions of CNTs aid in defining and routing electrical signals at sub-nanometre level. The nanoscale dimensions and quantum nature of CNT restrict the usage of semi-classical transport-based models like FS and MS model to study their resistive size effects. Zhou et al. used first-principles density functional methods and supercell approach to study the size effects of CNT and copper nanowires [73]. The copper's nanowire resistance seems to violate the semi-classical Ohm's law prediction of resistance below 60 nm and seems to possess a high value. However, a comparative quantum simulation study of copper nanowire with 40 nm × 40 nm radius and 40 × 40 CNT bundles (with 1 nm radius each) showed a substantial decrease of CNT resistance to 4 Ω vs 53 Ω for copper, highlighting the potential use of CNTs as future nanoscale interconnects [73]. Srivastava et al. critically analysed the performance of a minimum sized inverter circuit that drives four minimum sized inverters at load side with bundled CNTs and copper as interconnects. The simulations reveal that contact-engineered CNTs outperform conventional copper and show less propagation delay at local interconnect level for small technological nodes (<40 nm). However, bundled CNTs show superior performance when compared to copper, at long intermediate and global contact levels [74]. It is to be noted that the electronic transport properties of the CNT including the threshold voltage, carrier type and barrier height significantly depend on Ohmic vs Schottky contact employed [75]. The metal-CNT junction properties in turn depend on the contacting material's work function and wettability. The conductive metallic elements like titanium, chromium and iron with decent wettability show less contact resistance to CNT than other metals [76]. The nanoscale interconnect resistance of CNT depends on the aspect ratio, contact area, density and the type of CNTs employed [77–79]. It is of interest to synthesize/fabricate smaller diameter nanotubes to increase the tube density to achieve smaller resistance (similar to that of copper), high current-carrying capacity and small power dissipation. However, controlling the individual tube diameter is a challenging task and requires strict growth protocols for reliable and large-scale yield.

The advancements in material discovery and nanofabrication tools have augmented the synthesis of CNT-based composites to cater specific levels of interconnects. The filling of CNT into a copper matrix (1) enables to increase the ampacity and minimize the interconnect resistance, (2) eases process variability and (3) allows easy integration to existing technologies [80]. For example, Subramaniam et al. have developed a high-performance CNT-copper composite that combines the high ampacity of CNT (6×10^8 A/cm^2—100 times higher than copper), and high conductivity of copper (4.7×10^5 S/cm—similar to copper) can be an excellent candidate for future nanoscale interconnect. The high ampacity of the CNT-copper composite is attributed to decrease in diffusion coefficient by 10^4 to that of pristine copper, with CNT's resisting copper diffusion [63].

The layered two-dimensional (2D) material graphene, a monolayer of carbon atoms possessing similar properties to that of CNT, is considered as a viable candidate for nanoscale interconnects due to its atomic flat 2D nature that helps to design miniaturized large-scale integrated circuits [81,82]. Graphene being nano-sized also offers high thermal conductivity (around 3,100–5,300 W/m K), aids in better thermal management and self-heating issues that arise due to high current density in nanoscale interconnects [83,84]. Graphene nanoribbons (GNRs) considered as one-dimensional quantum confined graphene are considered as potential candidate to replace copper-based nanoscale interconnects [85]. The electronic transport in short GNRs ($L < 100$ nm) is hindered by contact junction and type, and shows negligible dependence on the grain boundaries, defects and roughness in edges [86]. It is to be noted that the resistive size effects in copper are predominantly dominated by surface and grain boundary scattering as per FS and MS model, respectively (see equations 3.1 and 3.2). GNRs have the potential to phase out this strong dependence and can evolve as reliable nanoscale interconnects. Wang et al. examined gate level (INV, NAND

and NOR) and 32-bit microprocessor system level architecture using 7 nm FinFET technology and observed an 8% increase in energy-delay-product over conventional copper and asserted the improvement is due to reduced parasitic capacitances of GNRs [87]. The theoretical resistivity of the GNRs can be made lower than copper by Fermi level tuning with appropriate doping and intercalation techniques [88]. The calculated resistivity using Landauer formalism (accounting edge scatterings and band-gap variations) for copper, tungsten and GNR (undoped and doped) is shown in Figure 3.9. It can be observed that the doped GNR's resistivity is less than copper for all interconnect linewidth less than 40 nm, and the corresponding size effect in GNR is less when compared to copper [89]. The use of advanced nanofabrication and surface passivation techniques will aid to achieve a perfect reflection surface corresponding to $p = 1$ in Figure 3.9a for experimental implementation. For example, Jiang et al. have experimentally observed that $FeCl_3$ intercalated doped multilayered GNRs exhibit smaller resistance with decreasing wire-widths (see dashed line and circular dots in Figure 3.9b). The comparison of $FeCl_3$ doped GNR and copper under similar conditions resulted in 20% improvement in estimated delay per unit length, 25% energy efficiency improvement at local interconnect level, 75% energy efficiency improvement level at global interconnect level and improved reliability over 12-nm-thin copper layers down to a width of 20 nm (see Figure 3.9b) [89].

It seems like carbon-based nanoscale interconnects can potentially replace the current copper-based interconnect technology to mitigate resistive size effects, improved thermal performance, reduced self-heating effects and achieve high conductivity and ampacity at nanoscale dimensions. However, the process variability and high growth temperature of CNTs are some of the major factors that hinder the integration and synthesis of large-scale reliable CNT nanoscale interconnects.

3.3.3 Topological Metals

The exponential increase in resistance of copper at nanoscale due to size effects can account for 35% of total signal delays and approximately 50% of dynamic power consumption in modern ICs [90]. There is an urgent need for energy-efficient

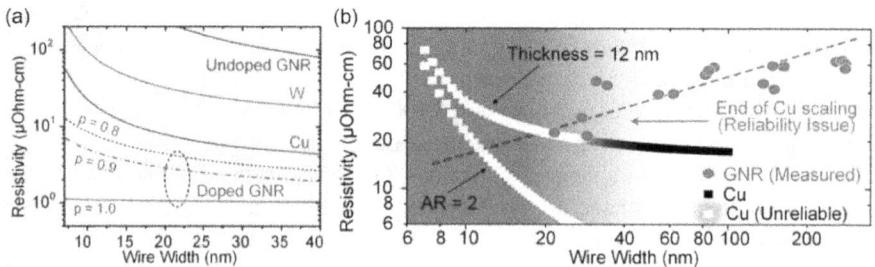

FIGURE 3.9   (a) The comparison of calculated resistivity of undoped and doped graphene nanoribbons vs conventional copper and tungsten interconnects at different wire-widths. (b) The comparison of experimental $FeCl_3$ doped GNR and copper at different wire-widths. The shaded region for wire-widths less than 40 nm highlights the copper scaling effects. Permission from J. Jiang et al., *Nano Lett.* 17, 1482–1488, 2017.

material that can compete or outperform conventional copper-based interconnect at nanoscale dimensions. Topological metals have garnered recent attention due to their surface-protected states, wherein the surface charge carriers do not undergo back-scattering, like conventional metals [91,92]. The absence of back-scattering minimizes the contribution of resistive size effects at nanoscale dimensions, thereby making them suitable candidates for nanoscale interconnects. Unlike conventional metals, the conduction band and valence band touch at discrete pairs of nodal points near the Fermi level in the Brillouin zone, giving rise to novel electronic properties [93–95]. The band crossings result in the occurrence of robust surface states connecting the nodal points to yield open Fermi arcs. The distinct band topology leads to both bulk and surface state conduction [96]. The individual contribution of surface vs bulk can be manipulated by controlling the thickness of the topological metal material. For instance, decrease in thickness leads to increase in surface conduction contribution that is void of any back-scattering leading to overall decrease in resistivity, which is in quite contrast to conventional copper metal resistive scaling with reduced dimensions [97].

The current research is focussed on discovering and understanding the structure-property relationships of topological metals to cater functional device and interconnect applications. The phenomenally low-resistance values of topological metals at reduced dimensions are comparable to conventional copper, which is an exciting reason to pursue them as future interconnects. The study of recent topological metals including transition pnictides (TaAs, NbAs, TaP, NbP), layered materials (WTe$_2$/MoTe$_2$), Cd$_3$As$_2$ and ZrTe$_5$ with high carrier density and low resistivity opens up new avenues for the design and development of future nanoscale interconnects [98–103]. Zheng et al. discovered stable NbAs nanobelts and synthesized thin samples (thickness <300 nm) and reported their experimental room-temperature resistivity is near and even less than that of bulk copper resistivity (see Figure 3.9). The resistivity of thin film samples is considerably less at all thicknesses, when compared to bulk NbAs, agreeing well with the theoretical Fermi-arc transport nature of the surface states (see Figure 3.10) [104].

The down-scaling of resistivity vs decreasing thickness is experimentally observed in ZrTe$_5$, a relatively new topological material. Niu et al. studied the electronic transport properties of ZrTe$_5$ thin films (thickness <20 nm) and observed their room-temperature resistivity is less than 10 μΩcm, close to conventional copper values [102]. Han et al. recently synthesized topological metal MoP nanowires using chemical vapour deposition method and reported MoP nanowire resistivity is less than conventional copper below 500 nm^2 cross-sectional area [105]. The low resistivity of nanoscale topological metals will be an important parameter that goes into design consideration of next generation of interconnects. However, their integration to current silicon-based technologies needs to be critically addressed for full-scale production. To address compatibility issues, researchers from IBM have identified CoSi, a Si-CMOS technology compatible topological metal that shows resistive down-scaling property with reduced dimensions [106]. Lin et al. have studied the CoSi using first-principles transport calculations and identified a critical thickness, below which surface states dominate and show resistive down-scaling property; and conversely show increase in resistance with reduced dimensions, like conventional

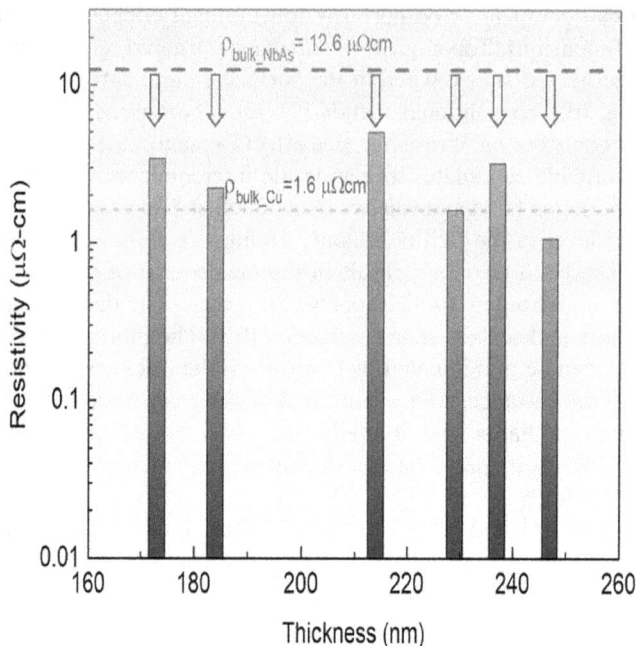

FIGURE 3.10 The scaling of measured room-temperature resistivity with thickness for the topological metal NbAs. The black dashed line represents the bulk resistivity data of the NbAs. The grey dashed line represents the bulk resistivity data of the conventional copper metal. The down arrow marks point to decrease of resistivity for all the nanobelt NbAs samples vs their bulk value. The experimental data of NbAs is taken from Ref. [104].

metals above the critical thickness [107]. Future research on comprehensive scaling of resistance in topological metals and advancements in low-temperature synthesis techniques to produce scalable nano-sized samples will facilitate an alternative solution to current copper-based interconnect challenges.

3.3.4 MXENES

Transition metal carbides, nitrides and carbonitrides are a new class of two-dimensional materials commonly referred as MXenes, with a generic formula $M_{n+1}X_nT_x$ (M = transition metal, X = carbon/oxygen and T_x = surface functional groups and n = 1, 2, 3 or 4) [108,109]. The potential to alter/substitute/dope different chemical compositions at M and X-sites, coupled with myriad surface terminations yield diverse stable material systems with unique properties that can be catered to specific applications. MXenes host several unique properties including high electrical conductivity and high mechanical strength, and the ability to bond to a variety of species makes them suitable candidate for nano-electronic, electromagnetic interference shielding and energy-storage applications [110–113]. Most of the synthesized MXenes are predicted to be conductive, although the conductivity type and nature (metallic vs semiconductor) can be altered by doping and surface functionalization [110].

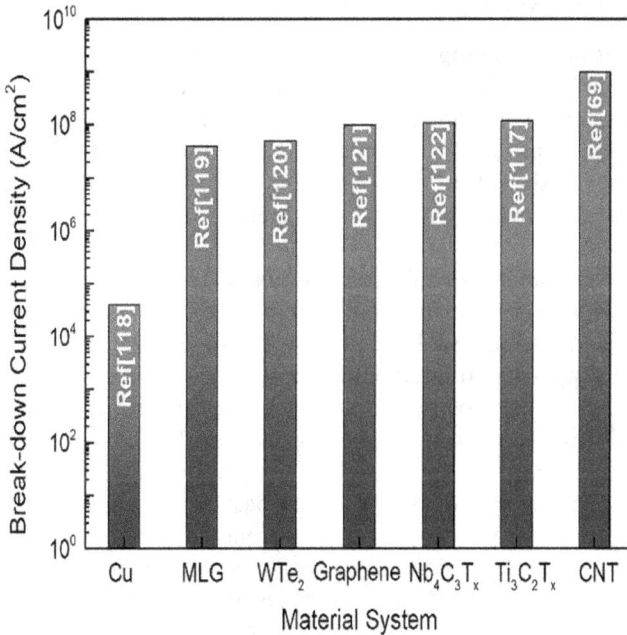

FIGURE 3.11 The comparison of break-down current density of future possible nanoscale interconnect material systems with conventional copper metal.

$Ti_3C_2T_x$ is the first discovered MXene material that has high electrical conductivity and Young's modulus of 4,600 S/cm and 330 GPa respectively, and is considered as a potential candidate to be used in high-current nanoscale electronic applications [114–116]. Lipatov et al. recently synthesized high-quality monolayer $Ti_3C_2T_x$ flakes (much like graphene) with high electrical conductivity of 11,000 S/cm and a break-down current density of 1.2×10^8 A/cm^2, which is comparable to CNTs and is almost twice that of copper element (see Figure 3.11) [117]. The ability to synthesize/fabricate sub-nanometre level flakes coupled with high break-down current density makes $Ti_3C_2T_x$ an ideal candidate for future nanoscale interconnects [118–121].

The experimental observation of high break-down current density in monolayer flakes of $Ti_3C_2T_x$ triggered researchers to design, develop and explore MXene-based material systems to figure out the optimum material choice that can compete or even outperform the reported value. Lipatov et al. have synthesized alternate promising $Nb_4C_3T_x$ flakes and observed a similar break-down current density of 1.1×10^8 A/cm^2 [122]. The research on MXene-based interconnects is still in its infancy level and future progress in areas of large-scale reliable synthesis techniques and ability to form stable composites that are compatible with silicon-based technologies would need to be tested. The comprehensive understanding of material chemistry and structure-property relationships in MXenes would possibly result in ultra-thin nanoscale materials that can withstand huge current densities for next-generation interconnects.

3.4 SUMMARY AND OUTLOOK

The continuous down-scaling of chip size is not only posing fundamental physical dimensional limitations to semiconductor channel materials, but equally poses a serious threat to interconnect materials that route the electric signals within an IC for its proper functioning. The currently widely used copper-based interconnect seriously suffers from resistive size effects, i.e., increase in resistance at nanoscale dimensions due to increased surface and grain boundary scattering effects.

There is an immediate need for the design and development of alternate material systems with long mean free path (to counter the nanoscale resistive scattering effects) and withstand high current densities at reduced dimensions to tackle current copper-based interconnect challenges. The advancements in material engineering and heterostructure engineering can help to synthesize defect-free materials with clean interfaces to support specular reflection to decrease the effect of resistive size effects in elemental copper-alternate metals like rhodium, ruthenium and cobalt. The nanoscale carbon-based materials including CNT, graphene and multilayer graphene offer high current-carrying capacities of the order of 10^{12} A/cm^2 at nanoscale dimensions which is an excellent candidate material system for future nanoscale interconnect systems. The topologically protected surface states of topological metals like NbAs prevent any back-scattering of charge carriers and show down-scaling of resistance with reduced dimensions, which needs to be further explored to understand the novel surface vs bulk physics, to be able to replace current copper-based interconnect technologies. The recent discovery of 2D MXene-based materials systems like $Ti_3C_2T_x$ with high break-down current density, comparable to that of CNT and twice that of conventional copper metal, seems to be a promising candidate for next-generation interconnect systems.

The alternate nanoscale interconnect material system's chemistry with interface liner/barrier layers needs to be thoroughly investigated to understand and design innovative planar vs vertical integrated novel architectures for efficient thermal management protocols. However, the copper-alternate interconnect material systems should be process-friendly, i.e., able to withstand different chemical and thermal treatments during IC fabrication process, environmentally benign and cost-effective to be considered as future reliable nanoscale interconnects.

REFERENCES

[1] Y. Chen, Z. Shu, S. Zhang, P. Zen, H. Liang, M. Zheng, and H. Duan, "Sub-10nm fabrication: Methods and applications," *Int. J. Extrem. Manuf.* 3, 032002 (2021).

[2] S. Wang, X. Liu, and P. Zhou, "The road of 2D semiconductors in the silicon age," *Adv. Mater.* 34, 206886 (2022).

[3] D. Akinwande, C. Huyghebaert, C.-H. Wang, M. I. Serna, S. Goossens, L.-J. Li, H-S. P. Wong, and F. H. L. Koppens, "Graphene and two-dimensional materials for silicon technology," *Nature* 573, 507–518 (2019).

[4] T. Chowdhury, E. C. Sadler, and T. J. Kempa, "Progress and prospects in transition-metal dichalcogenide research beyond 2D," *Chem. Rev.* 120, 12563–12591 (2020).

[5] W. Choi, N. Choudhary, G. H. Han, J. Park, D. Akinwande, and Y. H. Lee, "Recent development of two-dimensional transition metal dichalcogenides and their applications," *Mater. Today* 20, 116–130 (2017).

[6] H. Kim, and H. N. Alshareef, "Mxetronics: Mxene-enabled electronic and photonic devices," *ACS Mater. Lett.* 2, 55–70 (2020).

[7] S. Guha, A. Kabiraj, and S. Mahapatra, "High-throughput design of functional-engineered MXene transistors with low-resistance contacts," *NPJ Comput. Mater.* 8, 202 (2022).

[8] H. Liu, A. T. Neal, Z. Zhu, Z. Luo, X. Xu, D. Tomanek, and P. D. Ye, "Phosphorene: An unexplored 2D semiconductor with a high hole mobility," *ACS Nano* 8, 4033–4041 (2014).

[9] S. Das, W. Zhang, L. R. Thoutam, Z. Xiao, A. Hoffmann, M. Damarteau, and A. Roelofs, "A small signal amplifier based on ionic liquid gated black phosphorus field effect transistor," *IEEE Electron Device Lett.* 36, 621–623 (2015).

[10] M. A. Kharadi, G. F. A. Malik, F. A. Khanday, K. A. Shah, S. Mittal, and B. K. Kaushik, "Review-silicene: From material to device applications," *J. Solid State Sci. Technol.* 9, 115031 (2020).

[11] ITRS, *International Technology Roadmap for Semiconductors 2.0*, Interconnect, 2015.

[12] J. S. Chawla, and D. Gall, "Specular electron scattering at single-crystal Cu(001) surfaces," *Appl. Phys. Lett.* 94, 252101 (2009).

[13] F. Chen, and D. Gardner, "Influence of line dimensions on the resistance of Cu interconnections," *IEEE Electron Device Lett.* 19, 508–510 (1998).

[14] R. L. Graham, G. B. Alers, T. Mountsier, N. Shamma, S. Dhuey, S. Cabrini, R. H. Geiss, D. T. Read, and S. Peddeti, "Resistivity dominated by surface scattering in sub-50nm Cu wires," *Appl. Phys. Lett.* 96, 042116 (2010).

[15] D. Adams, and T. L. Alford, "Encapsulated silver for integrated circuit metallization," *Mater. Sci. Eng. R Rep.* 40, 207–250 (2003).

[16] D. Gall, "The search for the most conductive metal for narrow interconnect lines," *J. Appl. Phys.* 127, 050901 (2020).

[17] C. J. Uchibori, X. Zhang, P. S. Ho, and T. Nakamura, "Effect of chip-package interaction on mechanical reliability of Cu interconnects for 65 nm technology node and beyond," In: *2006 International Interconnect Technology Conference*, Burlingame, CA, pp. 196–198 (2006). doi: 10.1109/IITC.2006.1648686.

[18] H. Zahedmanesh, and K. Vanstreels, "Mechanical integrity of nano-interconnects; the impact of metallization density," *Micro Nano Eng.* 2, 35–40 (2019).

[19] V. Senez, T. Hoffman, P. L. Duc, and F. Murray, "Mechanical analysis of interconnected structures using process simulation," *J. Appl. Phys.* 93, 6039 (2003).

[20] T.-Y. Chiang, B. Sheih, and K. C. Saraswat, "Impact of Joule heating on scaling of deep sub-micron Cu/low-k interconnects," In: *2002 Symposium on VLSI Technology. Digest of Technical Papers (Cat. No.01CH37303)*, Honolulu, HI, pp. 38–39 (2002). doi: 10.1109/VLSIT.2002.101537.

[21] D. Save, F. Braud, J. Torres, F. Binder, C. Muller, J. O. Weidner, and W. Hasse, "Electromigration resistance of copper interconnects" *Microelectron. Eng.* 33, 75–84 (1997).

[22] "Copper interconnects: The evolution of microprocessors," *IBM Icons of Progress*. Last accessed from https://www.ibm.com/ibm/history/ibm100/us/en/icons/copperchip/.

[23] J. Gambino, "Process technology for copper interconnects," In: K. Seshan (ed), *Handbook of Thin Film Deposition*, third edition, William Andrew Publishing, Norwich, pp. 212–269 (2012).

[24] E. Milosevic, and D. Gall, "Copper Interconnects: Surface state engineering to facilitate specular electron scattering," *IEEE Trans. Electron Devic.* 66, 2692–2698 (2019).

[25] K. Fuchs, and N. F. Mott, "The conductivity of thin metallic films according to the electron theory of metals," *Math. Proc. Cambridge Philos. Soc.* 34, 100 (1938).

[26] E. H. Sondheimer, "The mean free path of electrons in metals," *Adv. Phys.* 1, 1–42 (1952).

[27] P. Kapur, J. P. McVittie, and K. C. Saraswat, "Technology and reliability constrained future copper interconnects," *IEEE Trans. Electron Devic.* 49, 590–597 (2002).

[28] T. S. Kuan, C. K. Inoki, G. S. Oehrlein, K. Rose, Y.-P. Zhao, G.-C. Wang, S. M. Rossnagel, and C. Cabral, "Fabrication and performance limits of sub-0.1μm Cu interconnects," *Mater. Res. Soc. Symp. Proc.* 612, 711 (2000).

[29] S. M. Rossnagel, and T. S. Kaun, "Alteration of Cu conductivity in the size effect regime," *J. Vac. Sci. Technol. B* 22, 240 (2004).

[30] T. Zhan, K. Sahara, H. Takeuchi, R. Yokogawa, K. Oda, Z. Jin, S. Deng, M. Tomita, Y-J. Wu, Y. Xu, T. Matsuki, H. Wang, M. Song, S. Guan, A. Ogura, and T. Watanabe, "Modification and characterization of interfacial bonding for thermal management of ruthenium interconnects in next-generation of very-large-scale integration circuits," *ACS Appl. Mater. Interf.* 14, 7392–7404 (2022).

[31] K. Banerjee, and A. Mehrotra, Global (interconnect) warming. *IEEE Circuit Devices Mag.* 17, 16–32 (2002).

[32] T. Zhan, K. Oda, S. Ma, M. Tomita, Z. Jin, H. Takezawa, K. Mesaki, Y-J Wu, Y. Xu, T. Matsukawa, T. Matsuki, and T. Watanabe, "Effect of thermal boundary resistance between the interconnect metal and dielectric interlayer on temperature increase of interconnects in deeply scaled VLSI," *ACS Appl. Mater. Interf.* 12, 22347–22356 (2020).

[33] C.-L. Lo, M. Catalano, A. Khosravi, W. Ge, Y. Ji, D. Y. Zemlyanov, L. Wang, R. Addou, Y. Liu, R. M. Wallace, M. J. Kim, and Z. Chen, "Enhancing interconnect reliability and performance by converting tantalum to 2D layered tantalum sulfide at low temperature," *Adv. Mater.* 31, 1902397 (2019).

[34] T. E. Swartz, and R. O. Pohl, "Thermal boundary resistance," *Reve. Mod. Phys.* 61, 605–668 (1989).

[35] H. Bishara, S. Lee. T. Brink, M. Ghidelli, and D. Dehm "Understanding grain boundary electrical resistivity in Cu: The effect of boundary structure," *ACS Nano* 15, 16607–116615 (2021).

[36] A. F. Mayadas, and M. Shatzkes, "Electrical resistivity model for polycrystalline films: The case of arbitrary reflection at external surfaces," *Phys. Rev. B.* 1, 1382 (1970).

[37] T. Sun, "Classical size effect in copper thin films: Impact of surface and grain boundary scattering on resistivity," *Electron. Theses Dissert.* 3927 (2009). (Last accessed at https://stars.library.ucf.edu/etd/3927).

[38] M. Cesar, D. Liu, D. Gall, and H. Guo, "Calculated resistances of single grain boundaries in copper," *Phys. Rev. Appl.* 2, 044007 (2014).

[39] T.-H. Kim, X.-G. Zhang, D. M. Nicholson, B. M. Evans, N. S. Kulkarni, B. Radhakrishnan, E. A. Kenik, and A.-P. Li, "Large discrete resistance jump at grain boundary in copper nanowire," *Nano. Lett.* 10, 3096–3010 (2010).

[40] Y. Kitaoka, T. Toni, S. Yoshimoto, T. Hirahara, S. Hasegawa, and T. Ohba, "Direct detection of grain boundary scattering in damascene Cu wires by nanoscale four-point probe resistance measurements," *Appl. Phys. Lett.* 95, 0521110 (2009).

[41] M. Cesar, D. Gall, and H. Guo, "Reducing grain-boundary resistivity of copper nanowires by doping," *Phy. Rev. Appl.* 5, 054018 (2016).

[42] International Roadmap for Devices and Systems, 2022 Update, More Moore. Last accessed at https://irds.ieee.org/images/files/pdf/2022/2022IRDS_MM.pdf.

[43] K. Croes, Ch. Adelmann, C. J. Wilson, H. Zahedmanesh, O. V. Pedreira, C. Wu, A. Lesniewska, H. Oprins, S. Beyne, I. Ciofi, D. Kocaayy, M. Stucchi, and Z. Tokei, "Interconnect metals beyond copper: Reliability challenges and opportunities," In:

2018 IEEE International Electron Devices Meeting (IEDM), San Francisco, CA, pp. 5.3.1–5.3.14 (2018). doi: 10.1109/IEDM.2018.8614695.

[44] M. R. Baklanov, C. Adelmann, L. Zhao, and S. De Gendt, "Advanced interconnects: Materials, processing and reliability," *ECS J. Solid State Sci. Technol.* 4, Y1–Y4 (2015).

[45] D. Gall, "Electron mean free path in elemental metals," *J. Appl. Phys.* 119, 085101 (2016).

[46] Melting point data is taken from American Elements. https://americanelements.com/meltingpoint.html.

[47] J. R. Lloyd, J. Clemens, and R. Snede, "Copper metallization reliability," *Microelectron. Reliab.* 39 1595–1602 (1999).

[48] A. Jog, T. Zhou, and D. Gall, "Resistivity size effect in epitaxial Rh(001) and Rh(111) layers," *IEEE Trans. Electron Devic.* 68, 257–263 (2021).

[49] N. A. Laznillo, and D. C. Edelstein, "Reliability and resistance projections for rhodium and iridium interconnects from first-principles," *J. Vac. Sci. Technol. B* 40, 052801 (2022).

[50] A. Jog, and D. Gall, "Resistivity size effect in epitaxial iridium layers," *J. Appl. Phys.* 130, 115103 (2021).

[51] V. Founta, J. P. Soulie, K. Sankaran, K. Vanstreeels, K. Opsomer, P. Morin, P. Lagrain, A. Franquet, D. Vanhaeren, T. Conard, J. Meersschaut, C. Detavernier, J. V. de Vondel, I. D. Wolf, G. Pourtois, Z. Toskei, J. Swerts, and C. Adelmann, "Properties of ultrathin molybdenum films for interconnect applications," *Materialia* 24, 1010511 (2022).

[52] L. G. Wen, P. Roussel, O. V. Pedreira, B. Briggs, B. Groven, S. Dutta, M. I. Popovici, N. Heylen, I. Ciofi, K. Vanstreels, F. W. Osterberg, O. Hansen, D. H. Petersen, K. Opsomer, C. Detavernie, C. J. Wilson, S. V. Elshocht, K. Croes, J. Bömmels, Z. Tőkei, and C. Adelmann, "Atomic layer deposition of ruthenium with tin interface for sub-10 nm advanced interconnects beyond copper," *ACS Appl. Mater. Interf.* 8, 26119–26125 (2016).

[53] D. Y. Kim, J.-S. NA, C.-S. Lai, R. Humayun, and M. Danek, "Depositing ruthenium layers in interconnect metallization," US10731250B2 US Patent (2018).

[54] M. F. J. Vos, S. N. Chopra, M. A. Verheijen, J. G. Ekerdt, S. Agarwal, W. M. M. Kessels, and J. M. Mackus, "Area-selective deposition of ruthenium by combining atomic layer deposition and selective etching," *Chem. Mater.* 31, 3878–3882 (2019).

[55] E. Milosevic, S. Kersongpanya, and D. Gall, "The resistivity size effect in epitaxial Ru (0001) and Co (0001) layers," In: *2018 IEEE Nanotechnology Symposium (ANTS)*, Albany, NY, pp. 1–5 (2018). doi: 10.1109/NANOTECH.2018.8653560.

[56] D. Choi, "Potential of ruthenium and cobalt as next-generation semiconductor interconnects," *Korean J. Met. Mater.* 56, 605–610 (2018).

[57] S. Dutta, S. Kundu, A. Gupta, G. Jamieson, J. F. G. Granados, J. Bommels, C. J. Wilson, Z. Tokei, and C. Adelmann, "Highly scaled ruthenium interconnects," *IEEE Electron Device Lett.* 38, 949–951 (2017).

[58] S. Ilani, and P. L. McEuen, "Electron transport in carbon nanotubes," *Annu. Rev. Cond. Mater. Phys.* 1, 1–25 (2010).

[59] J.-C. Charlier, X. Blasé, and S. Roche, "Electronic and transport properties of nanotubes," *Rev. Mod. Phys.* 79, 677 (2007).

[60] M. J. Allen, V. C. Tung, and R. B. Kaner, "Honeycomb carbon: A review of graphene," *Chem. Rev.* 110, 132–145 (2010).

[61] P. Avouris, and F. Xia, "Graphene applications in electronics and photonics," *MRS Bull.* 37, 1225–1234 (2012).

[62] ITRS International Technology working Groups. *International Technology Roadmap for Semiconductors*, ITRS, 2010.

[63] C. Subramaniam, T. Yamada, K. Kobashi, A. Sekiguchi, D. N. Futaba, M. Yumura, and K. Hata, "One-hundred-fold increase in current carrying capacity in a carbon nanotube-copper composite," *Nat. Commun.* 4, 2202 (2013).

[64] P. Poncharal, C. Berger, Y. Yi, Z. L. Wang, and W. A. de. Heer, "Room temperature ballistic conduction in carbon nanotubes," *J. Phys. Chem. B* 106, 12104–12118 (2002).

[65] A. S. Mayorov, R. V. Gorbachev, S. V. Morozov, L. Britnell, R. Jalil, L. A. Ponomarenko, P. Blake, K. S. Novoselvo, K. Watanabe, T. Taniguchi, and A. K. Geim, "Micrometer-scale ballistic transport in encapsulated graphene at room temperature," *Nano. Lett.* 11, 2396–2399 (2011).

[66] S. Chen, B. Shan, Y. Yang, G. Yuan, S. Huang, X. Lu, Y. Zhang, Y. Fu, L. Ye, and J. Liu, "An overview of carbon nanotubes bases interconnects for microelectronic packaging," In: *2017 IMAPS Nordic Conference on Microelectronics Packaging (NordPac)*, Gothenburg, Sweden, pp. 113–119 (2017). doi: 10.1109/NORDPAC.2017.7993175.

[67] P. Kim, L. Shi, A. Majumdar, and P. L. McEuen, "Thermal transport measurements of individual multiwalled nanotubes," *Phys. Rev. Lett.* 87, 215502 (2001).

[68] A. Javey, J. Guo, Q. Wang, M. Lundstrom, and H. Dai, "Ballistic carbon nanotube field-effect transistors," *Nature* 424, 654–657 (2003).

[69] Z. Yao, C. L. Kane, and C. Dekker, "High-field electrical transport in single-wall carbon nanotubes," *Phys. Rev. Lett.* 84, 2941 (2000).

[70] P. Goli, H. Ning, X. Li, C. Y. Lu, K. S. Novoselov, and A. A. Balandin, "Thermal properties of graphene-copper-graphene heterogenous films," *Nano Lett.* 14, 1497–1503 (2014).

[71] F. Kreupl, A. P. Graham, G. S. Duesberg, W. Steinhogl, M. Liebau, E. Unger, and W. Honlein, "Carbon nanotubes in interconnect applications," Microelectronic Engineering, 64, 399–408, (2002).

[72] A. Eatemadi, H. Daraee, H. Karimkhanloo, M. Kouhi, N. Zarghami, A. Akbarzadeh, M. Abasi, Y. Hanifehpour, and S. W. Joo, "Carbon nanotubes: Properties, synthesis, purification, and medical applications," *Nanoscale Res. Lett.* 9, 334 (2014).

[73] Y. Zhou, S. Sreekala, P. M. Ajayan, and S. K. Nayak, "Resistance of copper nanowires and comparison with carbon nanotube bundles for interconnect applications using first principles calculations," *J. Phys. Condens. Matter* 20, 095209 (2008).

[74] N. Srivastava, and K. Banerjee, "Performance analysis of carbon nanotube interconnects for VLSI applications," In: *ICCAD-2005. IEEE/ACM International Conference on Computer-Aided Design*, San Jose, CA, pp. 383–390 (2005). doi: 10.1109/ICCAD.2005.1560098.

[75] Z. Chen, J. Appenzeller, J. Knoch, Y. Lin, and P. Avouris, "The role of metal-nanotube contact in the performance of carbon nanotube field-effect transistors," *Nano. Lett.* 5, 1497–1502 (2005).

[76] S. C. Lim, J. H. Jang, D. J. Bae, G. H. Han, S. Lee, I.-S. Yeo, and Y. H. Lee, "Contact resistance between metal and carbon nanotube interconnects: Effect of work function and wettability," *Appl. Phys. Lett.* 95, 264103 (2009).

[77] C. Lan, D. N. Zakharov, and R. G. Reifenberger, "Determining the optimal contact length for a metal/multiwalled carbon nanotube interconnect," *Appl. Phys. Lett.* 99, 213112 (2008).

[78] Y. Awano, S. Sato, D. Kondo, M. Ohfuti, A. Kawabata, M. Nihei, and N. Yokoyama, "Carbon nanotube via interconnect technologies: Size-classified catalyst nanoparticles and low-resistance ohmic contact information," *Phys. Stat. Sol. A* 203, 3611 (2006).

[79] M. K. Majumder, J. Kumar, V. R. Kumar, and B. K. Kaushik, "Performance analysis for randomly distributed mixed carbon nanotube bundle interconnects," *Micro Nano Lett.* 9, 792–796 (2014).

[80] B. Uhlig, J. Liang, J. Lee, R. Ramos, A. Dhavamani, N. Nagy, J. Dijon, H. Okuna, D. Kalita, V. Georgiev, A. Asenov, S. Amoroso, L. Wang, C. Millar, F. Konemann, B. Gotsmann, G. Goncalves, B. Chen, R. R. Pandey, R. Chen, and A. Todri-Sanial, "Progress on carbon nanotube BEOL interconnects," In: *2018 Design, Automation and Test in Europe Conference and Exhibition (DATE)*, Dresden, Germany, pp. 937–942, (2018). doi: 10.23919/DATE.2018.8342144.

[81] X. Du, I. Skakchko, A. Barker, and E. Andrei, "Approaching ballistic transport in suspended graphene," *Nat. Nanotechnol.* 3, 491–495 (2008).

[82] C. Lee, X. Wei, J. W. Kysar, and J. Hone, "Measurement of the elastic properties and intrinsic strength of monolayer graphene," *Science* 321, 385–388 (2008).

[83] A. A. Balandin, S. Ghosh, W. Bao, I. Calizo, D. Teweldebrhan, F. Miao, and C. N. Lau, "Superior thermal conductivity of single-layer graphene," *Nano Lett.* 8, 902–907 (2008).

[84] S. Ghosh, I. Calizo, D. Teweldebrhan, E. P. Pokatilov, D. L. Nika, A. A. Balandin, W. Bao, F. Miao, and C. M. Lau, "Extremely high thermal conductivity of graphene: Prospects for thermal management application in nanoelectronic circuits," *Appl. Phys. Lett.* 92, 151911 (2008).

[85] H. Wang, H. S. Wang, C. Ma, L. Chen, C. Jiang, C. Chen, X. Xie, A.-P. Li, and X. Wang, "Graphene nanoribbons for quantum electronics," *Nat. Rev. Phys.* 3, 791–802 (2021).

[86] A. Behnam, A. S. Lyons, M.-H. Bae, E. K. Chow, S. Islam, C. M. Neumann, and E. Pop, "Transport in nanoribbon interconnects obtained from graphene grown by chemical vapor deposition," *Nano Lett.* 12, 4424–4430 (2012).

[87] N. C. Wang, S. Sinha, B. Cline, C. D. English, G. Yeric, and E. Pop, "Replacing copper interconnects with graphene at 7-nm node," In: *2017 IEEE International Interconnect Technology Conference (IITC)*, Hsinchu, Taiwan, pp. 1–3 (2017). doi: 10.1109/IITC-AMC.2017.7968949.

[88] C. Xu, H. Li, and K. Banerjee, "Modeling, analysis, and design of graphene nanoribbon interconnects," *IEEE Trans. Electron Devic.* 56, 1567–1568 (2009).

[89] J. Kiang, J. Kang, J. Cao, W. Xie, H. Zheng, J. H. Chu, W. Liu, and K. Banerjee, "Intercalation doped multilayer-graphene-nanoribbons for next-generation interconnects," *Nano Lett.* 17, 1482–1488 (2017).

[90] A. Ceyhan, M. Jung, S. Panth, S. K. Lim, and A. Naeemi, "Evaluating chip-level impact of Cu/low-*k* performance degradation on circuit performance at future technology nodes," *IEEE Trans. Electron Devic.* 62, 940–946 (2015).

[91] P. Liu, J. R. Williams, and J. J. Cha, "Topological nanomaterials," *Nat. Rev. Mater.* 4, 479–496 (2019).

[92] A. A. Burkov, "Topological semimetals," *Nat. Mater.* 15, 1145–1148 (2016).

[93] N. P. Armitage, E. J. Mele, and A. Vishwanath, "Weyl and Dirac semimetals in three-dimensional solids," *Rev. Mod. Phys.* 90, 015001 (2018).

[94] B. Yan and C. Felser, "Topological materials: Weyl semimetals," *Annu. Rev. Condens. Matter Phys.* 8, 337–354 (2017).

[95] L. R. Thoutam, M. Tangi, and S. M. Shivaprasad, "Novel emerging materials: Introduction and evolution," In: L. R. Thoutam, S. Tayal, J. Ajayan (eds), *Emerging Materials*, Springer, Singapore, pp. 3–36 (2022).

[96] J. Hu, S.-Y. Su, N. Ni, and Z. Mao, "Transport of topological semimetals," *Ann. Rev. Mater. Res.* 49, 207–252 (2019).

[97] P. O. Sukhachov, M. V. Rakov, O. M. Teslyk, and E. V. Gorbar, "Fermi arcs and DC transport in nanowires of Dirac and Weyl semimetals," *Ann. Phys.* 532, 1900449 (2020) .

[98] D. Grassano, O. Pulci, A. M. Conte, and F. Bechstedt, "Validity of Weyl fermion picture for transition metals monopnictides TaAs, TaP, NbAs and NbP from ab initio studies," *Sci. Rep.* 8, 3534 (2018).

[99] Y. Sun, S.-C. Wu, and B. Yan, "Topological surface states and Fermi arcs of the noncentrosymmetric Weyl semimetals TaAs, TaP, NbAs, and NbP," *Phys. Rev. B* 92, 115428 (2015).

[100] L. R. Thoutam, Y. L. Wang, Z. L. Xiao, S. Das, A. Luican-Mayer, R. Divan, G. W. Crabtree, and W. K. Kwok, "Temperature dependent three-dimensional anisotropy of the magnetoresistance of WTe2," *Phys. Rev. Lett.* 115, 046602 (2015).

[101] P. Li, Y. Wen, X. He, Q. Zhang, C. Xia, Z.-M. Yu, S. A. Zhang, Z. Zhu, H. N. Alshareef, and X.-X. Zhang, "Evidence for topological type-II Weyl semimetal WTe_2," *Nat. Commun.* 8, 2150 (2017).

[102] J. Niu, J. Wang, Z. He, C. Zhang, X. Li, T. Cai, X. Ma, S. Jia, D. Yu, and X. Wu, "Electrical transport in nanothick $ZrTe_5$ sheets: From three to two dimensions," *Phys. Rev. B* 95, 035420 (2017).

[103] T. Liang, Q. Gibson, M. N. Ali, M. Liu, R. J. Cava, and N. P. Ong, "Ultrahigh mobility and giant magnetoresistance in the Dirac semimetal Cd_3As_2," *Nat. Mater.* 14, 280–284 (2015).

[104] C. Zhang, Z. Ni, J. Zhang, X. Yuan, Y. Liu, Y. Zou, Z. Liao, Y. Du, A. Narayan, H. Zhang, T. Gu, X. Zhu, L. Pi, S. Sanvito, X. Han, J. Zou, Y. Shi, X. Wan, S. Y. Savrasov, and F. Xiu, "Ultrahigh conductivity in weyl semimetal NbAs nanobelts," *Nat. Mater.* 18, 482–488 (2019).

[105] H. J. Han, S. Kumar, X. Ji, J. L. Hart, G. Jin, D. J. Hynek, Q. P. Sam, V. Hasse, C. Felser, D. G. Cahill, R. Sundararaman, and J. J. Cha, "Topological metal MoP nanowire for interconnect," *Adv. Mater.* 35, 2208965 (2023).

[106] C.-T. Chen, U. Bajpai, N. A. Lanzillo, C.-H. Hsu, H. Lin, and G. Liang, "Topological semimetals for scaled back-end-of-line interconnect beyond Cu," In: *2020 IEEE International Electron Devices Meeting (IEDM)*, San Francisco, CA, pp. 32.4.1–32.4.4 (2020). doi: 10.1109/IEDM13553.2020.9371996.

[107] S.-W. Lien, I. Garate, U. Bajpai, C.-Y. Huang, C.-H. Hsu, Y.-H. Tu, N. A. Lanzillo, A. Bansil, T.-R. Chang, G. Liang, H. Lin, and C.-T. Chen, "Unconventional resistivity scaling in topological semimetal CoSi." *npj Quantum Mater.* 8, 3 (2023). doi: 10.1038/s41535-022-00535-6.

[108] Y. Gogotsi, and B. Anasori, "The rise of MXenes," *ACS Nano* 13, 8491–8494 (2019)

[109] Z. Fu, N. Wang, D. Legut, C. Si, Q. Zhang, S. Du, T. C. Germann, J. S. Francisco, and R. Zhang, "Rational design of flexible two-dimensional MXenes with multiple functionalities," *Chem. Rev.* 119, 11980–12031 (2019).

[110] A. Champagne, and J.-C. Charlier, "Physical properties of 2D MXenes: From a theoretical perspective," *J. Phys. Mater.* 3, 032006 (2021).

[111] H. Kim, and H. N. Alshareef, "MXetronics: MXene-enabled electronics and photonic devices," *ACS Mater. Lett.* 2, 55–70 (2020).

[112] F. Shahzad, M. Alhabeb, C. B Hatter, B. Anasori, S. ManHong, C. M. Koo, and Y. Gogotsi, "Electromagnetic interference shielding with 2D transition metal carbides (MXenes)," *Science* 353, 1137–1140 (353).

[113] B. Anasori, M. R. Lukatskaya, and Y. Gogotsi, "2D metal carbides and nitrides (MXenes) for energy storage," *Nat. Rev. Mater.* 2, 16098 (2017).

[114] M. Naguib, M. Kurotglu, W. Presser, J. Lu, J. Niu, M. Heon, L. Hultman, Y. Gogotsi, and M. W. Barsoum, "Two-dimensional nanocrystals produced by exfoliation of Ti_3AlC_2," 23, 4248–4253 (2011).

[115] A. Lipatov, M. Alhabeb, M. R. Luukatskaya, A. Boson, Y. Gogotsi, and A. Sinitskii, "Effect of synthesis on quality, electronic properties and environmental stability of individual monolayer Ti_3C_2 MXene flakes," *Adv. Electron. Mater.* 2, 1600255 (2016).

[116] A. Lipatov, H. Lu, M. Alhabeb, B. Anasori, A. Gruverman, Y. Gogotsi, and A. Sinitskii, "Elastic properties of 2D Ti_3C_2Tx Mxene monolayers and bilayers," *Sci. Adv.* 4, eaat0491 (2018).

[117] A. Lipatov, A Goad, M. J. Loes, N. S. Borobeva, J. Abourahma, Y. Gogotsi, and A. Sinitskii, "High electrical conductivity and breakdown current density of individual monolayer Ti$_3$C$_2$Tx MXene flakes," *Matter* 4, 1413–1427 (2021).

[118] M. E. T. Molares, E. M. Hohberger, Ch. Schaeflein, R. H. Blick, R. Neumann, and C. Trautmann, "Electrical characterization of electrochemically grown single copper nanowires," *Appl. Phys. Lett.* 82, 2139 (2003).

[119] K.-J. Lee, A. P. Chandrakasan, and J. Kong, "Breakdown current density of CVD-grown multilayer graphene interconnects," *IEEE Electron Device Lett.* 32, 557–559 (2011).

[120] M. J. Mleczko, R. L. Xu, K. Okabe, H.-H. Kuo, I. R. Fisher, H.-S. P. Wong, Y. Nishi, and E. Pop, "High current density and low thermal conductivity of atomically thin semi-metallic WTe$_2$," *ACS Nano* 10, 7507–7514 (2016).

[121] R. Murali, Y. Yang, K. Brenner, T. Beck, and J. D. Meindl, "Breakdown current density of graphene nanoribbons," *Appl. Phys. Lett.* 94, 243114 (2009).

[122] A. Lipatov, M. J. Loes, N. S. Vorobeva, S. Bagheri, J. Abourahma, H. Chen, X. Hong, Y. Gogotsi, and A. Sinitskii, "High breakdown current density in monolayer Nb$_4$C$_3$T$_x$ MXene," *ACS Mater. Lett.* 3, 1088–1084 (2021).

4 Analysis of Simultaneous Switching Noise and IR Drop

Debaprasad Das, Mounika Bandi,
Sandip Bhattacharya, and Subhajit Das

4.1 INTRODUCTION

With the enormous improvements in interconnect technology, the International Technology Roadmap for Semiconductors (ITRS) is encouraging to replace traditional copper (Cu) interconnects by carbon nanomaterials to alleviate the durability and reliability problems of Cu for next generation on chip interconnects. In nanometer dimensions, the properties of the bulk nanomaterials cannot be applied and the devices must be analyzed in atomic level. The molecular devices like organic transistors, memory and bulk molecular logic, carbon nanotubes, mono-molecular transistors, and graphene nanoribbon are being explored as emerging technologies. CNTs have high melting point, each carbon atom is joined to three other carbon atoms to form a strong covalent bond. This leaves each carbon atom with a spare electron, which forms delocalized electrons within the tube, i.e., carbon nanotubes, can conduct electricity. The applicability of metallic CNTs as interconnect has been studied in [1,2] extensively, and it is proposed as the replacement for copper-based interconnects. Replacement of copper interconnects with carbon-based conductors such as carbon nanotube (CNT) and graphene nanoribbon (GNR) materials reduces the weight of interconnects by up to 90%. Several researchers have modeled and performed analyses to investigate the pertinence of carbon nanotube and graphene nanoribbon over Cu interconnects as VLSI interconnect systems. CNT and GNR interconnects offer game-changing design possibilities over traditional copper-based power distribution networks (PDN). Power integrity is an important analysis in VLSI physical design flow to check the desired voltage and current levels from the driver to the load. Today, power integrity plays an important role in the success and failure of the electronic products [3,4]. Though there are many research works on modeling of CNT and GNR for signal integrity [1,5], there are only a few studies on power integrity for IR drop and simultaneous switching noise [11,13]. The copper-based PDNs suffer power supply voltage drop and simultaneous switching noise issues due to resistance, inductance, and capacitance of interconnects. This causes variations in the power supply voltage and ground, which result in delay, causing logic failure. In this chapter, we present the IR drop and simultaneous switching noise analysis in CNT and GNR interconnects and performance comparison is made w.r.t. Cu-based

DOI: 10.1201/9781003331650-4

interconnects. For analyses, different ITRS technology nodes [6] are considered. It is found that the IR drop in CNT and GNR power interconnects is significantly less as compared to copper-based PDN. The impact of IR drop on timing delay is greatly reduced in CNT- and GNR-based power network in comparison to that of copper interconnects [7–13].

4.2 MODELING OF CNT AND GNR INTERCONNECTS

4.2.1 Equivalent Circuit Parameters of SWCNT

4.2.1.1 Resistance of SWCNT Interconnect

The resistance of a SWCNT can be modeled by following parts: (1) contact resistance R_C, (2) quantum resistance R_Q, and (3) ohmic resistance R_O. Total resistance is expressed as

$$R_{cnt} = R_C + R_Q + R_O \tag{4.1}$$

$$R_{cnt} = R_C + R_Q \tag{4.2}$$

where l_{CNT}=length of SWCNT and λ=mean free path (MFP) of electron.
The quantum and ohmic resistances are given by

$$R_Q = \frac{h}{4e^2} \tag{4.3}$$

$$R_O = R_Q \left(\frac{l_{CNT}}{\lambda} \right)$$

and respectively, where h=Plank's constant and e=electronic charge. The contact resistance R_C can be as high as 100 kΩ.

4.2.1.2 Inductance of SWCNT Interconnect

Single-walled CNT has kinetic inductance (L_K) in addition to magnetic inductance (L_M). The magnetic inductance can be expressed as

$$L_M = \left(\frac{\mu}{2\pi} \right) \cdot ln \left(\frac{ht}{d} \right) \tag{4.4}$$

where ht=height of the carbon nanotube from ground plane and d=diameter of the CNT. The kinetic inductance is given by

$$L_K = \frac{h}{2e^2 \times v_F} \tag{4.5}$$

where v_F = Fermi velocity, which is usually taken as 8×10^5 m/s for CNT. As there are four conducting channels in single-walled carbon nanotube, the effective kinetic inductance is $\frac{L_K}{4}$. It is empirical that the magnetic inductance is irrelevant as compared to the kinetic inductance. Therefore, the inductance of SWCNT can be written as $\frac{L_K}{4} \approx L_{CNT}$.

4.2.1.3   Capacitance of SWCNT Interconnect

The capacitance of SWCNT is modeled by two parts: (1) quantum capacitance C_Q and (2) electrostatic capacitance C_E. The expression for quantum capacitance is given by

$$C_Q = \frac{2e^2}{h \times v_F} \tag{4.6}$$

The electrostatic capacitance is given by

$$C_E = \frac{2\pi \in}{ln\left(\dfrac{ht}{d}\right)} \tag{4.7}$$

The effective quantum capacitance due to the four conducting channels in SWCNT is $4C_Q$.

4.2.2   RLC PARAMETERS OF A SWCNT BUNDLE

Nano interconnects are proposed using a bundle of SWCNT due to the large intrinsic resistance (6.45 kΩ) of SWCNT. Regardless, chirality of CNTs has lack of control, a bundle consists of both metallic and semiconducting CNTs. It is found that the fraction of metallic CNTs P_m is one-third in a bundle. The rest of the CNTs are semiconducting in a bundle and they do not participate in current conduction. If w and t are the width and thickness of interconnects, respectively, the number of single-walled carbon nanotubes along the x and z directions can be expressed as

$$n_x = \frac{(w-d)}{x} \tag{4.8}$$

and

$$n_z = \frac{2(w-d)}{\sqrt{3}x} + 1 \tag{4.9}$$

where d = diameter of CNT and x = inter-CNT distance. We assume both densely packed ($P_m = 1$) and sparsely packed ($P_m = \frac{1}{3}$) CNT bundles in this work. For the densely packed bundle, the inter-CNT distance is d and for the sparsely packed bundle it is $\sqrt{3}d$. Using the expressions of n_x and n_z, the total number of CNTs in a bundle can be expressed as

$$n_{CNT} = n_x n_z - \frac{n_z}{2} \quad \text{if } n_z \text{ is even} \tag{4.10}$$

and

$$n_{CNT} = n_x n_z - \frac{n_z - 1}{2} \quad \text{if } n_z \text{ is odd} \tag{4.11}$$

Using a number of CNTs, the bundle is formed in parallel, the effective resistance, inductance, and capacitance are given by parallel combination of resistance, inductance, and capacitance of the individual CNTs. Therefore, the resistance, inductance, and capacitance of a CNT bundle of length l_{CNT} can be expressed as follows.

4.2.2.1 Resistance of SWCNT Bundle

$$R_b = \frac{R_{CNT}}{n_{CNT}} \tag{4.12}$$

4.2.2.2 Inductance of SWCNT Bundle

$$L_b = \frac{l_{CNT} L_{CNT}}{n_{CNT}} \tag{4.13}$$

4.2.2.3 Capacitance of SWCNT Bundle

$$C_b = l_{CNT} \frac{C_Q^b C_E^b}{C_Q^b + C_E^b} \tag{4.14}$$

where the quantum capacitance (C_Q^b) and electrostatic capacitance (C_E^b) of the SWCNT bundle are given by

$$C_Q^b = C_Q^{CNT} \cdot n_{CNT} \tag{4.15}$$

$$C_E^b = 2C_{En} + \frac{(n_x - 2)}{2} C_{Ef} + \frac{3(n_z - 2)}{5} C_{En} \tag{4.16}$$

where C_{En} = capacitance, calculated from the ground plane at a distance equal to spacing s between the interconnects and capacitance = C_{Ef} is calculated assuming ground plane at a distance equal to spacing plus width of interconnects $s + w$.

4.2.3 Equivalent RLC Parameters of MWCNT

A multi-walled CNT has a number of shells, each having different conducting channels depending on its diameter. Hence, the total number of conducting channels in a MWCNT is given by the sum of the conducting channels N_i of all the shells.

$$N = \sum_{i=1}^{p} [N_i] \tag{4.17}$$

The resistance of a shell consists of three parts: quantum resistance R_Q, scattering-induced ohmic resistance R_O, and imperfect contact resistance R_{mc}.

4.2.3.1 Intrinsic Resistance R_{CNT}

$$R_{CNT} = R_Q + R_O.L = \left(\frac{h}{2e^2 N} \right).\left(1 + \frac{L}{\lambda} \right) \tag{4.18}$$

where $h/2e^2$ is 12.9 kΩ, and L, λ, and N are the length, mean free path (MFP), and number of conducting channels of the MWCNT, respectively. The contact resistance R_{mc} can range from 0 to 100 kΩ for different growth processes. R_{mc} in MWCNT could be very small compared to the total resistance.

4.2.3.2 Inductance of MWCNT

The inductance of MWCNT has two parts: (1) magnetic inductance and (2) kinetic inductance. The kinetic inductances per unit length of a shell are given by

$$L_{k/channel} = \frac{h}{2e^2 \times v_F}.\frac{1}{2} \approx 8nH/\mu m \tag{4.19}$$

$$L_{k/shell} = \frac{L_{k/channel}}{N_{shell}(D)} \tag{4.20}$$

We have considered only the kinetic inductance and neglected the magnetic inductance as it is insignificant as compared to the kinetic inductance values.

4.2.3.3 Capacitance of MWCNT

The capacitance of a MWCNT also has two parts: (1) quantum capacitance C_Q and (2) electrostatic capacitance C_E.

The quantum capacitance per unit length of a shell can be expressed as

$$C_{Q/channel} = \frac{4e^2}{h \times v_F} \approx 193\,aF/\mu m \tag{4.21}$$

$$C_{Q/shell} = C_{Q/channel}\,N_{shell}(D) \tag{4.22}$$

The shell-to-shell capacitance per unit length C_S can be identified by using the coaxial capacitance formula as

$$C_S = \frac{2\pi\epsilon}{ln\left(D_{max}/D_{min}\right)} = \frac{2\pi\epsilon}{ln\left[D_{max}/(D_{min}-2\delta)\right]} \tag{4.23}$$

where D_{max} and D_{min} are the diameters of the outer and inner shells of adjacent coaxial shells and δ equals 0.34 nm.

4.2.4　EQUIVALENT CIRCUIT RLC PARAMETERS OF GNR

The number of graphene layers is given by

$$N_{layer} = 1 + Integer\left[\frac{t}{\delta}\right] \tag{4.24}$$

GNR interconnect is modeled by a distributed RLC network, where R_C = resistance due to imperfect contacts, R_Q = quantum resistance, and R_S ($=\frac{R_Q}{\lambda}$) = scattering resistance per unit length where λ = mean free path of electrons in GNR. The quantum resistance is defined as

$$R_Q = \frac{h/2e^2}{N_{ch}N_{layer}} = \frac{12.94k\Omega}{N_{ch}N_{layer}} \tag{4.25}$$

where N_{ch} = number of conducting channels (modes) in one layer, N_{layer} = number of GNR layers, h = Plank's constant ($=6.626\times10^{-34}$ J s), and e = electronic charge ($=1.6\times10^{-19}$ C).

We assumed $\lambda = 1$ μm in this work. A perfect contact resistance ($R_C = 0$ Ω) is assumed. The per unit length quantum capacitance can be expressed as

$$C_Q = N_{ch}N_{layer}\frac{4e^2}{h\times v_F} = N_{ch}N_{layer}\times193.18\,aF/μm \tag{4.26}$$

where $v_F = 8\times10^5$ m/s for GNR. The per unit length kinetic inductance can be expressed as

$$L_k = \frac{h/4e^2v_F}{N_{ch}N_{layer}} = \frac{8.0884}{N_{ch}N_{layer}}\,nH/μm \tag{4.27}$$

The number of conducting channels in single-layer graphene is given by

$$N_{ch} = \sum_{j=1}^{n_C}\left[1+e^{\frac{\left(E_{j,n}-E_F\right)}{k_BT}}\right]^{-1} + \sum_{j=1}^{n_v}\left[1+e^{\frac{\left(E_F-E_{j,h}\right)}{k_BT}}\right]^{-1} \tag{4.28}$$

where $j = 1, 2, 3,...$ is a positive integer, E_F is Fermi energy, k_B is the Boltzmann's constant, T is absolute temperature, and n_c and n_v are the number of conduction and valence subbands, respectively. $E_{j.n}$ and $E_{j.h}$ are the energy of electron and hole in jth subband as given by

$$E_j = \frac{jhv_F}{2W} \qquad (4.29)$$

where width $= W$, $j = 1, 2, 3,...$ is a positive integer. The number of conducting channels N_{ch} is 17, 12, 8, and 6, respectively, for metallic GNR of width 45, 32, 22, and 16 nm technology nodes for $E_F = 0.3$ eV.

4.3   MODELING OF POWER DISTRIBUTION NETWORK

A typical integrated circuit has single power and ground pins that are located generally at the opposite corners of an integrated circuit. Figure 4.1 shows a schematic of small integrated circuit containing 100 standard cells arranged in ten rows and ten columns. In this scenario, the standard cells that are placed near the power pin get proper power supply voltage level. However, the standard cells that are away from the power pin get a degraded power supply voltage. Similarly, the cells that are close to the ground pin get proper ground voltage and the cells that are away from the ground pin experience a greater ground voltage level. Therefore, the cells with a lesser power supply voltage are impacted in terms of propagation delay. The propagation delay is increased for these cells and the cells become relatively slower.

The equivalent circuit for IR-drop analysis is shown in Figure 4.2. In order to analyze the performance of power supply voltage drop we have considered an inverter chain.

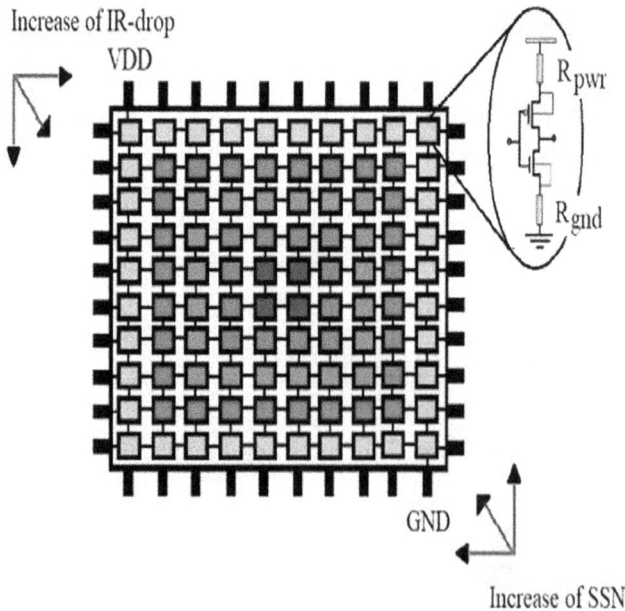

FIGURE 4.1   Block diagram of 10×10 standard cell-based integrated circuit.

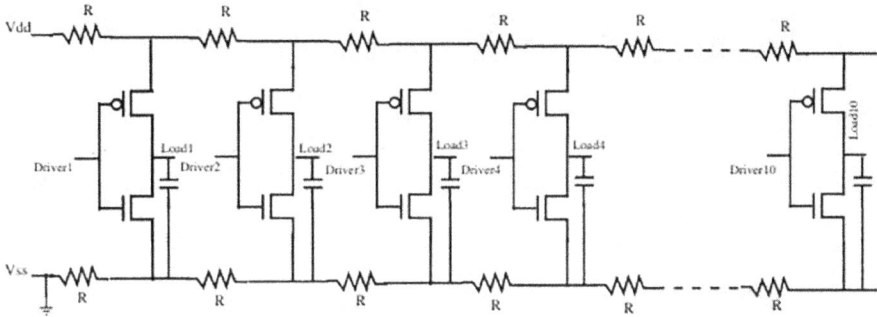

FIGURE 4.2 Schematic circuit for modeling power supply voltage drop where input power V_{dd} and ground V_{ss} are applied.

In this circuit diagram we have modeled the power V_{dd} and ground V_{ss} interconnects by series connected resistive networks to model each segment of power/ground interconnects. All the inverter cells are connected to same power V_{dd} and ground V_{ss}. The connection points are separated by an interconnect segment of length (L) that is varied from 1 to 100 μm.

The values of circuit parameters (RLC) are calculated using the analytical equations described in the previous section. The different lengths of interconnects are considered as (1, 5, 10, 20, 50, 100) μm. Due to the resistance on the PDN, the current results in a voltage drop across the power network and there is an increase in voltage across (also known as ground bounce) on the ground network. IR drop refers to the amount of decrease or increase in the power or ground voltage due to the resistance of power or ground interconnects [11].

In this work, we have used PDN consisting of ten inverters as standard logic cells placed at a uniform distance. The inverter cells are designed for different technology nodes. The cells drive a default load capacitance (C_i) of 0.1 pF, which represents the effective load that the particular cell is driving. A SPICE netlist of the circuit in Figure 4.1 is developed and simulated. The simulations are performed using the ITRS technology nodes at 45, 32, 22, and 16 nm nodes using Spice Simulator.

An induced voltage drop occurs in an IC PDN, when multiple output drivers switch simultaneously. Due to over-switching, it momentarily raises the ground voltage within the circuit relative to the system ground. This shift in the ground potential to a non-zero value is known as simultaneous switching noise (SSN) or ground bounce, which produces an oscillation on the output of a circuit. It translates into higher bit error at the receiver. This is more accurate for series network. Figure 4.3 represents the schematic diagram of SSN analysis considering ten number of stages was an analytical expression characterizing the SSN voltage presented on a parameter of RLC model where power V_{dd} and ground V_{ss} are connected. The output of power integrity is represented in Figure 4.4 as voltage waveforms at different points on the power and ground line where from the curve N1, N2, N3,..., N10 indicate the V_{dd} input voltage line, while N11, N22, N33,..., N111 show the V_{ss} ground voltage line. The V_{dd} values are 1.0, 0.9, 0.8 and 0.7 V for 45, 32, 22 and 16 nm different technology nodes, respectively.

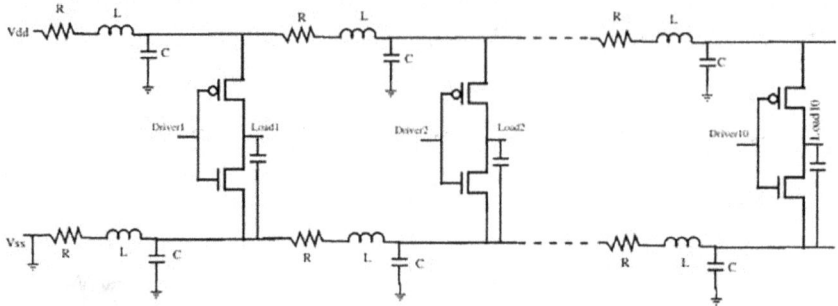

FIGURE 4.3   Schematic circuit for modeling equivalent parameters RLC for SSN where input power V_{dd} and ground V_{ss} are applied for nth number of load capacitance of 0.1 pF.

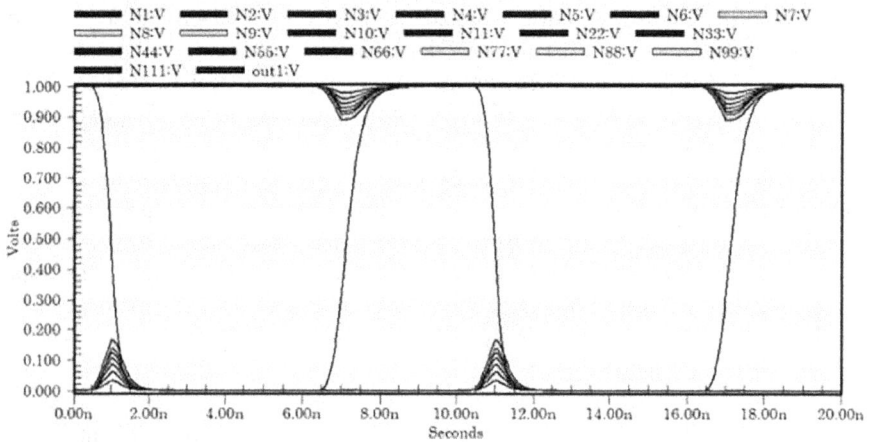

FIGURE 4.4   Voltage waveforms at different points on the power and ground line.

4.4   IR-DROP AND SSN ANALYSIS RESULTS

The IR drop analysis is performed for copper, carbon nanotube, and graphene nanoribbon-based interconnect systems for different lengths (1–100 µm) using ITRS technology nodes (16, 22, 32 and 45 nm) considering V_{dd} values are 1.0, 0.9, 0.8, and 0.7 V, respectively. The variation of IR drop with interconnect lengths is shown in Figure 4.5 at ten different stages of the inverter chain. It is observed that as the stage number increases the IR drop also increases. For 1 µm interconnect length, GNR shows least IR drop as compared to other interconnect systems.

It is also observed that for longer-length SWCNT, densely packed bundle has less IR drop as compared to MWCNT and Cu interconnect systems. The IR drop induced timing delay due to dynamic local power and ground different interconnect systems as shown in Figure 4.6. It is clear that for longer lengths, setup time violation occurs at a smaller stage number. It has also been observed that GNR interconnect exhibits lower delay compared to Cu, SWCNT, and MWCNT. Due to higher values of parasitic elements, the delay in copper is much higher compared to CNT and GNR

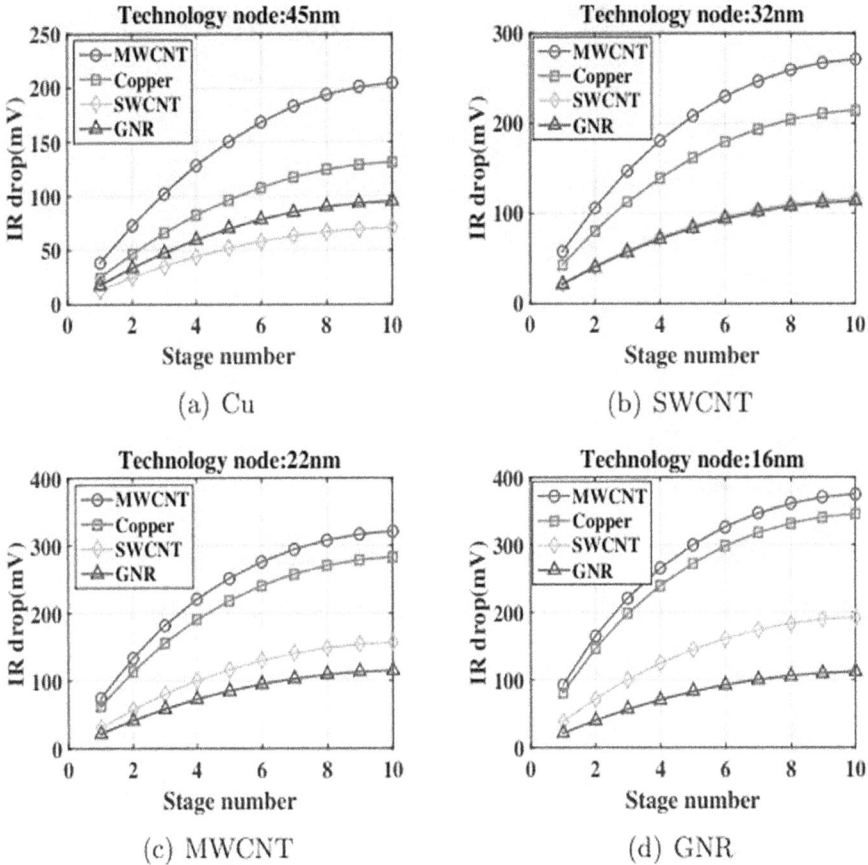

FIGURE 4.5 IR drop in local power and ground interconnects for Copper, SWCNT, MWCNT and GNR based PDNs. Each stage is separated by equal distance from the power and ground pad of distances of 10 μm are analyzed. The V_{dd} values are 1.0, 0.9, 0.8 and 0.7 V for 45, 32, 22 and 16 nm different technology nodes, respectively.

interconnects. Figure 4.7 shows due to over-switching it raises the ground voltage within the inverter chain network relative to the system ground. This shifting in the ground voltage is as SSN or ground bounce. SSN voltage vs stage number from here can be observed as the effect of switching for number of stages more in Cu-based interconnects than in CNT and GNR interconnects (Figure 4.7).

4.5 CONCLUSION

In this chapter, power integrity analysis is presented using CNT and GNR interconnects. It is observed that as the stage number increases the IR drop also

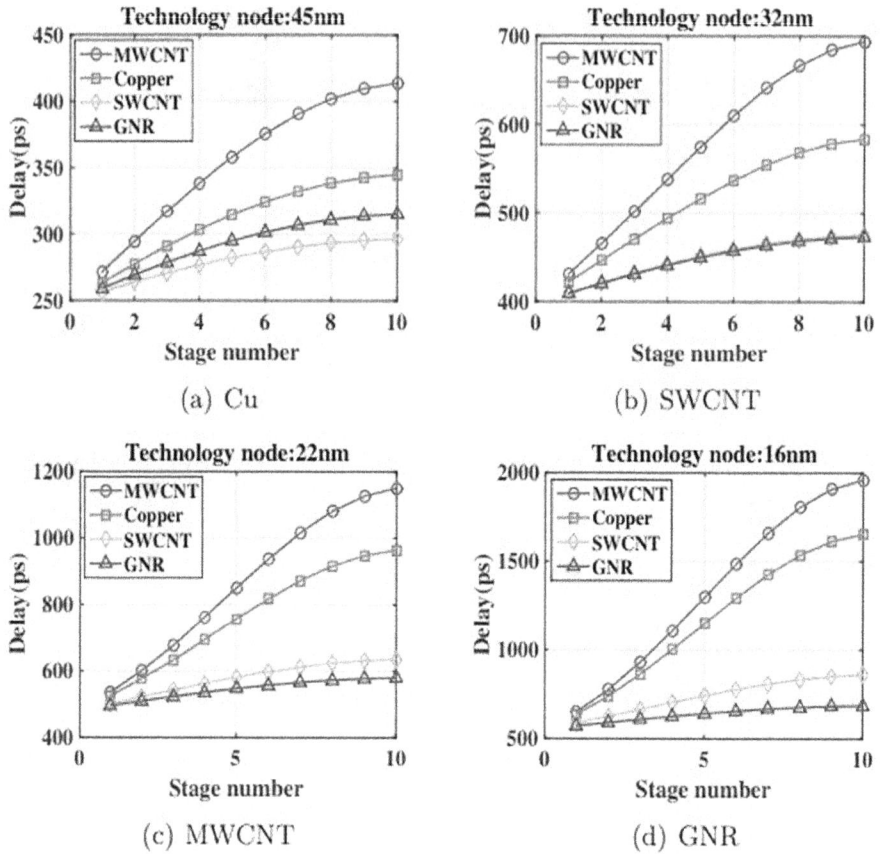

FIGURE 4.6   Timing delay due to dynamic IR drop in local power and ground interconnects for Copper, SWCNT, MWCNT and GNR based PDNs. Each stage is separated by equal distance from the power and ground pad of distances of 10 μm are analyzed of different technology nodes.

increases. GNR shows least IR drop as compared to other interconnect systems. It is investigated that for longer lengths setup time violation occurs at shorter stage number. It is concluded that GNR interconnect exhibits lower delay as compared to Cu, SWCNT, and MWCNT. The shifting in voltage on the ground lines occurs as SSN or ground bounce over the networks. We observed that the effect of switching for number of stages is more in Cu-based interconnects than CNT and GNR interconnects.

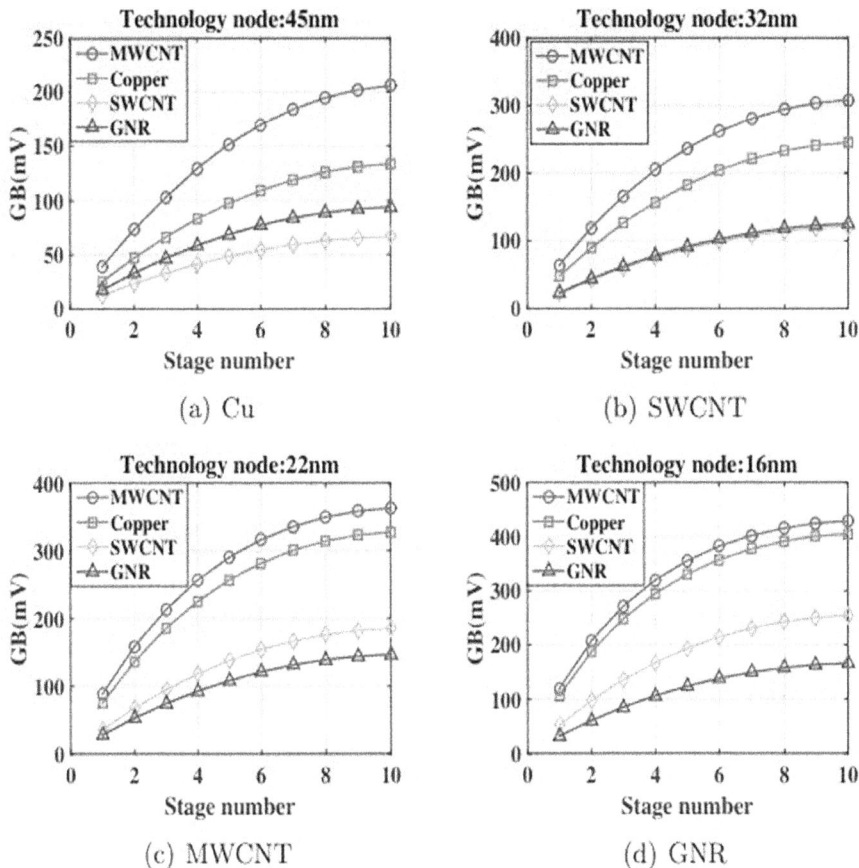

FIGURE 4.7 Simultaneous switching noise (SSN) or ground bounce in local power and ground interconnects for Copper, SWCNT, MWCNT and GNR based PDNs. Each stage is separated by equal distance from the power and ground pad of distances of 10 µm are analyzed of different technology nodes.

REFERENCES

[1] D. Das and H. Rahaman, "Analysis of Crosstalk in Single- and Multiwall Carbon Nanotube Interconnects and Its Impact on Gate Oxide Reliability," *IEEE Trans. Nanotechnol.* vol. 10, no. 6, pp. 1362–1370, 2011.

[2] M.K. Mazumder, B.K. Kaushik and S.K. Manhas, "Analysis of Delay and Dynamic Crosstalk in Bundled Carbon Nanotube Interconnects," *IEEE Trans. Electromag. Compat.* vol. 56. no. 6, pp. 1666–1673, 2014.

[3] N. Srivastava and K. Banerjee, "Performance Analysis of Carbon Nanotube Interconnects for VLSI Application," in *IEEE/ACM International Conference on Computer-Aided Design*, IEEE, pp. 383–390, 2005.

[4] C. Xu, H. Li and K. Banerjee, "Modelling, Analysis, and Design of Graphene Nano-Ribbon Interconnects," *IEEE Trans. Electron Devices* vol. 56, no. 8, pp. 1567–1578, 2009.

[5] H. Li, W. Yin, K. Banerjee, and J. Mao, "Circuit Modelling and Performance Analysis of Multi-Walled Carbon Nanotube Interconnects," *IEEE Trans. Electron Devices* vol. 55, no. 6, pp. 1328–1337, 2008.

[6] ITRS, *International Technology Roadmap for Semiconductors*, 2004, https://public. itrs.net.

[7] A. Giustiniani, V. Tucci and W. Zamboni, "Modelling Issues and Performance Analysis of High-Speed Interconnects Based on a Bundle of SWCNT," *IEEE Trans. Electron Devices* vol. 57, no. 8, pp. 1978–1986, 2010.

[8] N. Srivastava, H. Li, F. Kreupl and K. Banerjee, "On Applicability of Single-Walled Carbon Nanotubes as VLSI Interconnects," *IEEE Trans. Nanotechnol.* vol. 8, no. 4, pp. 542–559, 2009.

[9] D. Das and H. Rahaman, "Crosstalk and Gate Oxide Reliability Analysis in Graphene Nanoribbon Interconnects," in *2011 International Symposium on Electronic System Design*, IEEE, 2011, pp. 182–187.

[10] A. Naeemi and J.D. Meindl, "Design and Performance Modeling for Singlewalled Carbon Nanotubes as Local, Semi-Global, and Global Interconnects in Gigascale Integrated Systems," *IEEE Trans. Electron Devices* vol. 54, no. 1, pp. 26–37, 2007.

[11] K.T. Tang and E.G. Friedman, "Simultaneous Switching Noise in OnChip CMOS Power Distribution Networks," *IEEE Trans. VLSI Syst.* vol. 10, no. 4, pp. 487–493, 2002.

[12] D. Das and H. Rahaman," IR "Drop Analysis in Single-and Multi-Wall Carbon Nanotube Power Interconnnects in Sub-Nanometer Designs," in *2011 3rd Asia Symposium on Quality Electronic Design (ASQED)*, IEEE, 2011, pp. 174–183.

[13] D. Das and H. Rahaman, "Modeling of IR-Drop Induced Delay Fault in CNT and GNR Power Distribution Networks," in *2012 5th International Conference on Computers and Devices for Communication (CODEC)*, IEEE, 2012, pp. 1–4.

5 Temperature-Dependent RF Performance Analysis of GNR-Based Nano-Interconnect Systems

Santasri Giri Tunga, Hafizur Rahaman, and Subhajit Das

5.1 INTRODUCTION

Apprehensible attributes of graphene have been observed in its structure along with noticeable features in electrical, thermal, chemical, and optical properties [1–4]. The 2D form of carbon atoms which are patterned in a honeycomb manner is the base of monolayer graphene. Graphene nanoribbon (GNR) is stripped from this monolayer graphene whose properties are regulated by the edge chirality and width [5]. During GNR fabrication, controlled edge pattering aids GNR to manifest both metallic and semiconducting property. It has been observed that metallic GNR bears quasi-ballistic transport because of large MFP and thermal conductivity, high tensile and mechanical strength, and is mostly impervious to electromigration effect [6,7]. Multilayer GNR offers low resistance due to multiple conduction channels unlike single-layer GNR which owns high resistance [8,9]. These distinguishable features turn MLGNR to be a probable alternative to copper for interconnect application in the nano-scale industry.

Different research and studies have observed and explained the performance superiority of MLGNR over SLGNR [10,11]. Likewise, excellent behavioural performance is observed for pristine Side Contact MLGNR (SC MLGNR) over pristine Top Contact MLGNR (TC MLGNR) [12–14]. However, fabrication complexity of SC MLGNR motivated researchers to shift from SC MLGNR to TC MLGNR for chip-level interconnect application, and attempts have been made to make performance of TC MLGNR compatible with SC MLGNR. Pronounced developments have been noticed for TC MLGNR when it is intercalated with other dopant atoms and performance of the intercalation-doped TC MLGNR is found to supersede SC MLGNR in certain aspects [12–16].

DOI: 10.1201/9781003331650-5

Significant increase of resistivity of metal interconnects with rise in temperature results from higher scattering rate. Besides, scaling down of chip-level circuit causes an increase in current density that triggers Joule heating [17,18]. These outcomes of thermal variation steer degradation concerning reliability and RF performance as has been reported in [19–25].

In this work, examination of temperature-dependent RF performance is carried out for pristine TC and SC MLGNR along with different intercalation-doped TC MLGNR. Atoms of different dopants such as AsF_5, Li, and $MoCl_5$ are intercalated with TC MLGNR and the behavioural results are compared with pristine TC and SC MLGNR. Remarkable results are obtained for intercalation-doped TCMLGNR in terms of radiofrequency (RF) analysis. Temperature-dependent number of conducting channels and effective MFP are calculated as in [25] and results obtained are used in the present work for the calculation and analysis of radiofrequency behaviour.

5.2 EQUIVALENT RLC MODEL OF MLGNR INTERCONNECT

Two resistive elements namely lumped resistance (R_{Lumpd}) and distributed resistance (R_{Dist}) comprise the fundamental ESC model of MLGNR interconnect as in [26]. Alteration in interconnect length does not affect the lumped elements' performance, whereas pronounced effect is detected for distributed elements. Conducting channel-dependent quantum resistor R_{qt} and imperfect metal connecting layers contact resistor R_{ct} comprise the far and near terminal lumped resistance [12]. Thus, for MLGNR interconnect, the effective lumped resistance is the summation of quantum resistor and contact resistor given as [13]

$$R_{Lumpd} = {1} \Big/ {\sum (R_{qt} + R_{ct})^{-1}} \tag{5.1}$$

For MLGNR the effective distributed resistance (R_{efct}) is the aggregate of the resistances present in vertical and horizontal fragmented unit cells. The differential length of each unit cell along the length is considered as Δlen. Each segment comprises of the in-plane resistance along the horizontal fragmented segments (R_{hz}) and perpendicular conductance along the vertical layers (G_V) expressed mathematically as [12]

$$R_{hz} = \left(\frac{R_{qt}}{N_{ch} \Lambda_{efct}} \right) \Delta len \tag{5.2}$$

and

$$G_V = \left(\frac{W_d}{\rho_c \delta_s} \right) \Delta len \tag{5.3}$$

where δs, ρ_c, and Λ_{efct} are the interlayer spacing, the c-axis resistivity, and the effective MFP, respectively. E_f-dependent N_{ch} in each GNR layer and width (W_d) are obtained from the famous Fermi-Dirac equation [9]

$$N_{ch} = \sum_{b=1}^{nc} \left(1 + e^{\frac{E_f + E_b}{K_bT}} \right)^{-1} + \sum_{b=1}^{nv} \left(1 + e^{\frac{E_b - E_f}{K_bT}} \right)^{-1} \qquad (5.4)$$

where $j = 1, 2, 3,...$ a positive integer. K_b and T are Boltzmann's constant and temperature, respectively.

Ideally, in an interconnect, the conduction of current is contributed by all the layers of MLGNR. Kumar et al. [13] have calculated the maximum number of layers conducting current in MLGNR for various widths. However, as the layer number increases progressively, Fermi level is observed to decrease exponentially for each layer with respect to screening length, which limits the number of layers actively participating in conduction. Subsequently, the number of conducting layers decreases and effective resistance increases for an MLGNR [26]. As stated earlier, intercalation doping is an effectual way to enhance the performance of MLGNR due to increase in E_f, effective MFP, and interlayer distance. Consequently, in intercalation-doped TC MLGNR, the number of layers actively participating in current conduction increases with respect to its pristine counterpart.

5.3 GNR CAPACITANCE AND INDUCTANCE CALCULATION

The distributed elements in the MLGNR interconnect, along with the in-plane and vertical resistances, comprise of the effective inductance and effective capacitance. The inductive and capacitive impedance is determined as a function of frequency. Subsequently, the interconnect capacitance and inductance have significant contribution in determining the RF analysis. Nishad et al. calculated the capacitance as the aggregate of quantum capacitance and electrostatic capacitance [27,28]. In GNR, the quantum capacitance is a function of Fermi level-dependent number of conducting channels and is determined as [29]

$$\frac{1}{C_{qt}} = \frac{hv_f}{4e^2 N_{ch}} \qquad (5.5)$$

A schematic model to calculate the electrostatic capacitance (C_{ES}) is depicted in Figure 5.1. C_{ES} is calculated as a matrix of the interlayer capacitance (C_{Int}) and the capacitances found between the upper layer to ground (C_{UG}) and lower layer to ground (C_{LG}). The matrix is expressed as [25]

$$[C_{ES}] = \begin{bmatrix} C_{UG} + C_{Int} & -C_{Int} & 0 & 0 & 0 \\ -C_{Int} & 2C_{Int} & -C_{Int} & 0 & 0 \\ 0 & -C_{Int} & 2C_{Int} & -C_{Int} & 0 \\ 0 & -C_{Int} & 2C_{Int} & -C_{Int} & 0 \\ 0 & 0 & 0 & -C_{Int} & C_{UG} + C_{Int} \end{bmatrix} \qquad (5.6)$$

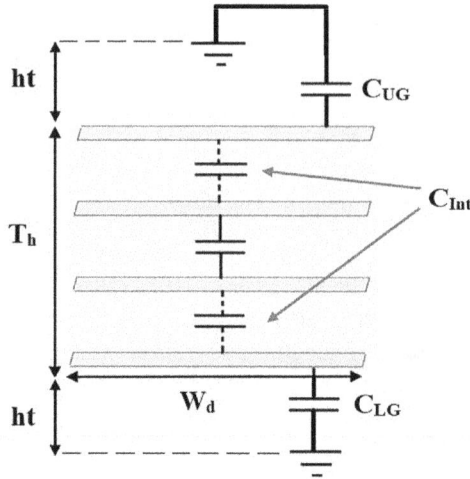

FIGURE 5.1   MLGNR interconnect capacitance model [31].

The capacitances C_{UG}, C_{LG}, and C_{Int} are calculated from [30] using conformal mapping method. In each fragmented segment of the interconnect, the product of the unit cell capacitance and the differential length gives the effective capacitance. Hence, for the length Δlen, the effective distributed capacitance of MLGNR interconnect is given as

$$\left(C_{efct}\right)^{-1} = \left(C_{Unit}\Delta len\right)^{-1} \tag{5.7}$$

Similarly, in each fragmented segment of the interconnect with length Δlen, the product of the unit cell inductance calculated as aggregate of kinetic and mutual inductance with the differential length gives the effective inductance as

$$\left(L_{efct}\right)^{-1} = \left(L_{Unit}\Delta len\right)^{-1} \tag{5.8}$$

For pristine MLGNR, E_f is considered as 0.2 eV, which shifts to higher value for different intercalation-doped MLGNR. Consequently, the MFP also enhances in inter-calation-doped MLGNR to that of pristine MLGNR. Significant change in MFP is observed for variation in temperature. Electron and phonon scattering is augmented with uprising temperature, which eventually drops down the MFP [32]. Effective MFP for analysis of radio frequency behaviour is obtained from [25].

5.4   FREQUENCY RESPONSE ANALYSIS OF MLGNR INTERCONNECT

The electro-thermal circuit model is demonstrated using multi-conductor-based methodology. Further, ABCD transmission matrix has been used for the investigation of RF performance of MLGNR interconnect [25]. An infinitesimal section of length Δlen from the total length of the interconnect is considered for the analysis. The calculation using ABCD parameters requires horizontal resistance (R_{hz}), effective inductance (L_{eff}), effective capacitance (C_{eff}), and the vertical conductance (G_v).

A matrix based on these variables has been formulated to obtain the transfer function of the transmission line.

A comprehensive combination of different elements for each network of the driver interconnect load (DIL) system is demonstrated in Figure 5.2 and using ABCD transmission matrix the open loop transfer function is derived [33,34]. An infinitesimal section of length Δlen of the MLGNR interconnect is considered for the analysis of temperature-dependent frequency response. An N_L layer MLGNR interconnect may be considered as a multi-conductor system which can be described using its resistance $[R]$, inductance $[L]$, capacitance $[C]$, and conductance $[G]$ parameters discussed above. Using these parameters, the per unit length impedance (Z_U) and admittance (Y_U) of the interconnect are defined as

$$Z_U = R_{Unit} + j\omega L_{Unit} \tag{5.9}$$

and

$$Y_U = G_{Unit} + j\omega C_{Unit} \tag{5.10}$$

For an unvarying RLC transmission line of length Δlen the ABCD parameter matrix is formulated as

$$\begin{bmatrix} A_{\Delta len} & B_{\Delta len} \\ C_{\Delta len} & D_{\Delta len} \end{bmatrix} = \begin{bmatrix} \cosh(\theta\Delta len) & Z_o\sinh(\theta\Delta len) \\ \left(\dfrac{1}{Z_o}\right)\sinh(\theta\Delta len) & \cosh(\theta\Delta len) \end{bmatrix} \tag{5.11}$$

where

$$Z_o = \sqrt{\dfrac{Z_{U(j\omega)}}{Y_{U(j\omega)}}} \tag{5.12}$$

and

$$\theta = \sqrt{Z_{U(j\omega)}Y_{U(j\omega)}} \tag{5.13}$$

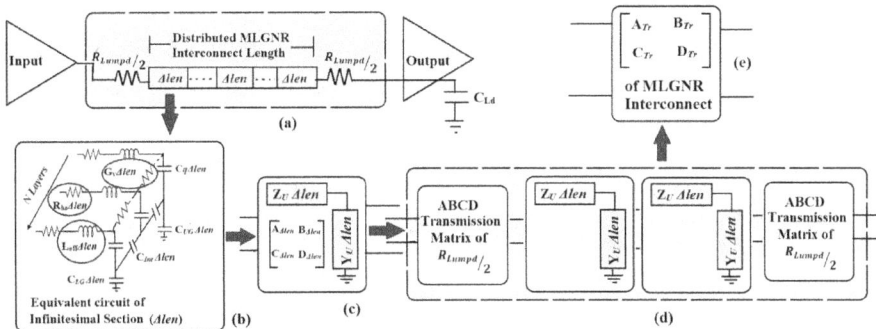

FIGURE 5.2 Equivalent RF model of MLGNR interconnect [25].

For the MLGNR interconnect system depicted in Figure 5.2, the overall ABCD transmission matrix may be explicitly expressed as

$$\begin{bmatrix} A_{Tr} & B_{Tr} \\ C_{Tr} & D_{Tr} \end{bmatrix} = \begin{bmatrix} 1 & \dfrac{R_{Lumpd}}{2} \\ 0 & 1 \end{bmatrix} \begin{bmatrix} \cosh(\theta\Delta len) & Z_o\sinh(\theta\Delta len) \\ \left(\dfrac{1}{Z_o}\right)\sinh(\theta\Delta len) & \cosh(\theta\Delta len) \end{bmatrix} \begin{bmatrix} 1 & \dfrac{R_{Lumpd}}{2} \\ 0 & 1 \end{bmatrix} \qquad (5.14)$$

From equation (5.14), we obtain the transfer function that determines the temperature-dependent unified frequency range. In this entire bandwidth, the signal can be restored fully without missing any information at the load. An analytical expression of the transfer function from which we can obtain the frequency range for both TC and SC MLGNR interconnect for capacitive load [17] of the DIL system shown in Figure 5.2 can be expressed as

$$H(j\omega) = \left(A_{Tr} + C_{Ld}B_{Tr}\right)^{-1} \qquad (5.15)$$

Figures 5.3 and 5.4 depict the frequency response of TC and SC MLGNR for the temperature values considered for the analysis and at 10, 50, and 100 µm interconnect length. For a temperature range of 233–423 K, the frequency response at 10 µm interconnect length shows that the 3 dB cut-off frequency for the TC MLGNR stretches from 5 to 15 MHz in contrast to that of 75 MHz to 2 GHz to its SC counterpart as is depicted in Figure 5.3.

Moreover, in Figure 5.4, for an interconnect length of 50 µm, the ranges of 3 dB cut-off frequency for TC and SC MLGNR are found to be 0.2–2.5 and 3–60 MHz, respectively. The frequency range for interconnect length of 100 µm in Figure 5.4 is found to be 50–520 KHz for TC MLGNR, while it is 10–100 MHz for SC MLGNR. For different interconnect lengths, it is observed that the bandwidth is more in SC MLGNR compared to that of TC MLGNR. In SC MLGNR, metal contact is connected to all the layers, consequently all layers participate in current conduction, which leads to notable drop in impedance. Unlike SC MLGNR, the current conduction in TC MLGNR is predominantly surface bound as metal contact is connected to the topmost layer. This results in high impedance which is attributed to the degradation in the RF performance of TC MLGNR. Further, it is also noticed that with increase in temperature, the cut-off frequency decreases irrespective of type of contact and length of interconnect because the resistance and capacitance present in the distributed and lumped elements of the interconnect are found to increase with increase in temperature. Besides, the impact of temperature-dependent scattering reduces the effective MFP at higher frequencies. Consequently, a significant drop both in operating frequency range and effective impedance occurs as is marked in the graphs of Figure 5.4.

The RF performance analysis of SC MLGNR is superior to its TC counterpart. However, complexity arises while fabricating SC MLGNR in planar technology. Enhancement of performance in TC MLGNR is accomplished with intercalation doping. In this method, dopant charge carriers are introduced in between the layers. The consequent results are

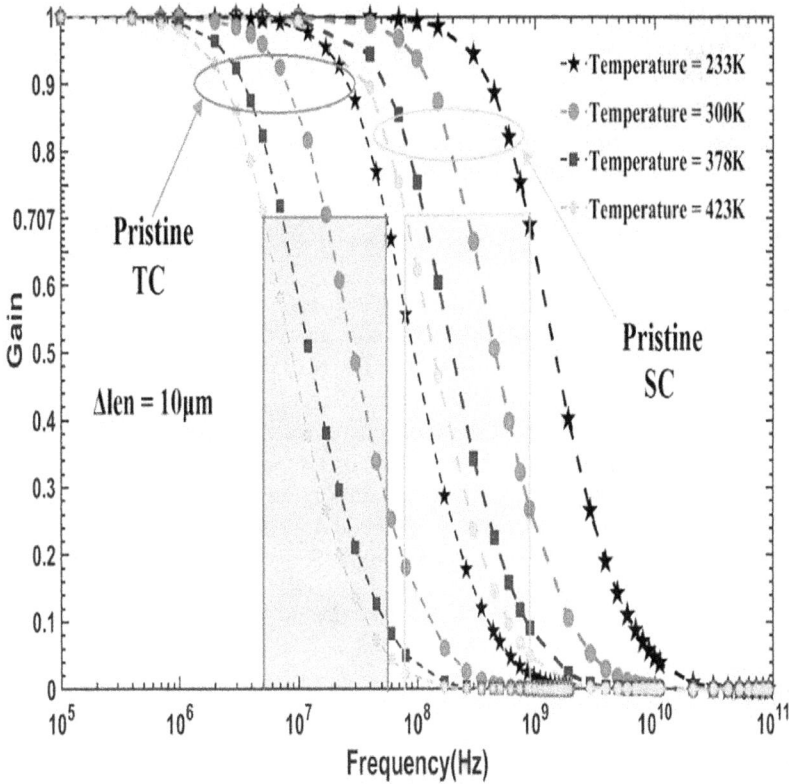

FIGURE 5.3 Temperature vs frequency response of pristine TC and SC MLGNR at 10 μm interconnect length.

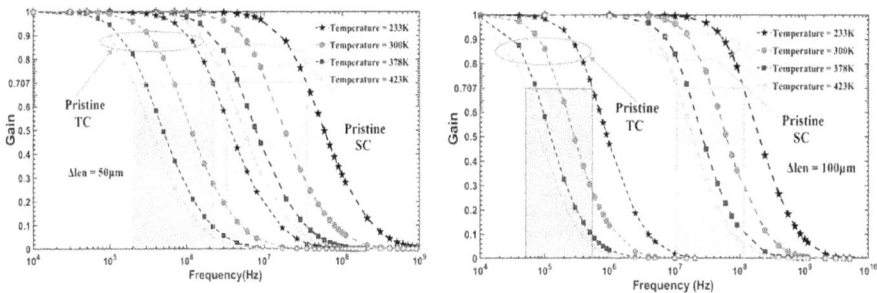

FIGURE 5.4 Temperature vs frequency response of pristine TC and SC MLGNR at an interconnect length of 50 and 100 μm.

upward shifting of Fermi level, expansion of the interlayer distance, lowered scattering phenomenon, higher MFP, and a considerable increase in conductivity.

The present work involves the study of frequency response for AsF_5, Li, and $MoCl_5$ intercalation-doped TC MLGNR. For a temperature range of 233–423 K and interconnect lengths of 10, 50, and 100 μm, the RF response for the intercalation-doped TC MLGNR is analysed. Similar to pristine TC MLGNR, with increase in temperature, the cut-off frequency for intercalation-doped MLGNR dropped; however, the bandwidth of frequency increased significantly with intercalation doping. It is noteworthy to mention that with intercalation doping, the increased Fermi level with electron density increased the total number of conducting channels for individual layer. Consequently, with intercalation, significant reduction of impedance is observed due to increased effective MFP and a substantial boost of the operating frequency is noticed.

Figure 5.5 demonstrates the 3 dB bandwidth for pristine TC and SC MLGNR along with intercalated TC MLGNR. Significant enhancement of bandwidth is observed for different intercalation-doped TC MLGNR compared to its pristine counterpart. The frequency range at different chip operating temperatures is represented in Table 5.1 obtained from the transfer function given in equation (5.15).

It is noteworthy to mention that among the different intercalation doping used for the RF analysis in this chapter, MoCl$_5$ intercalation-doped TCMLGNR has the highest Fermi level of 3.4 eV. Subsequently, the bandwidth is found to be larger than the other intercalants irrespective of the interconnect length and operating temperatures depicted in Figure 5.6.

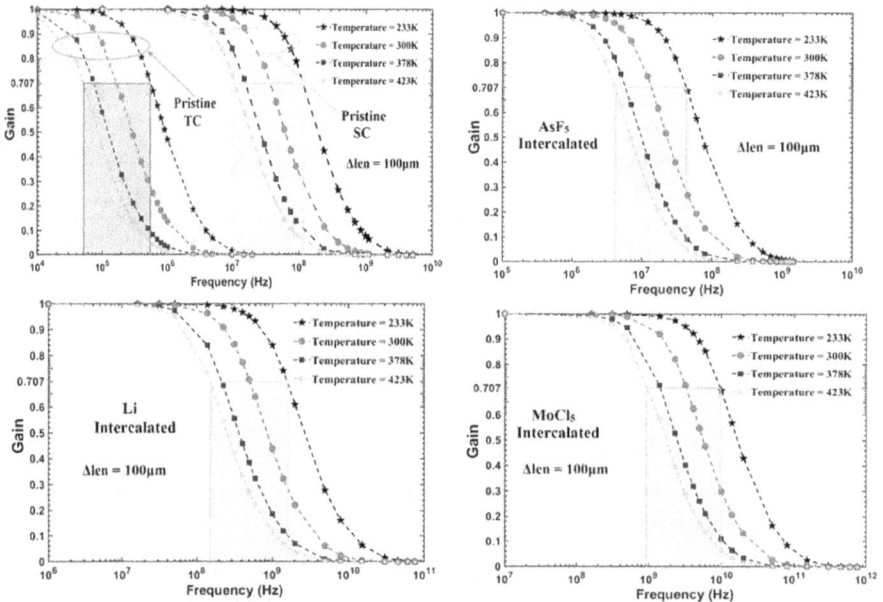

FIGURE 5.5 Temperature vs frequency response of pristine and different intercalation-doped TC MLGNR at an interconnect length of 100 μm.

TABLE 5.1

3 dB Bandwidth (MHz) for Pristine and Intercalated MLGNR and Copper for Interconnect Length = 100 µm ($W_d = 16$ nm, $p = 0.2$)

Temperature (K)	Pristine TC	Pristine SC	AsF$_5$ Intercalated	Li Intercalated	MoCl$_5$ Intercalated	Cu
233	7.1	103	54	1,100	9,100	171
300	2.4	34	22	530	3,200	130
378	1	11	7	220	1,050	101
423	0.7	10.2	4	170	900	86

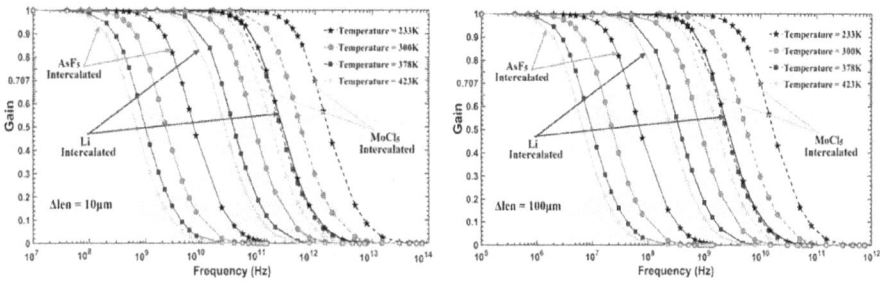

FIGURE 5.6 Temperature vs frequency response of AsF$_5$, Li, and MoCl$_5$ TC MLGNR at an interconnect length of 10 and 100 µm.

5.5 CONCLUSION

This RF analysis of pristine and intercalation doped TC- and SC-MLGNR nano-interconnect system has been performed with the variation of temperature. A comprehensive RF model based on ABCD transmission matrix is developed for the illustration of the temperature-dependent frequency response of pristine TC and SC MLGNR along with AsF$_5$, Li, and MoCl$_5$ intercalation-doped TC MLGNR. Considering the technology node of 16 nm, interconnect length and temperature are varied to obtain the electro-thermal RF analysis. Observation revealed that at higher temperature, the cut-off frequency is less compared to the cut-off frequency at lower temperatures, whereas for shorter interconnect length (10 µm), the operating bandwidth is in few GHz range than at longer interconnect lengths (100 µm) which is in few MHz range. Besides, the frequency response reflects that an effective approach to attain superior RF analysis is intercalation doping in MLGNR. In essence, this work furnishes a valuable insight for intercalation-doped MLGNR nano-interconnect application at high-frequency range.

REFERENCES

[1] Novoselov, K. S., Geim, A. K., Morozov, S. V., Jiang, D., Zhang, Y., Dubonos, S. V., Grigorieva, I. V., and Firsov, A. A.: Electric field effect in atomically thin carbon films, *Science* vol. 306, no. 5696, pp. 666–669, 2004.

[2] Mak, K. F., Lui, C. H., and Heinz, T. F.: Measurement of the thermal conductance of the graphene/SiO$_2$ interface, *Appl. Phys. Lett.* vol. 97, p. 221904, 2010.

[3] Avouris, P., Chen, Z., and Perebeinos, V.: Carbon based electronics, *Nat. Nanotechnol.* vol. 2, no. 10, pp. 605–615, 2007.

[4] Berger, C., Song, Z., Li, T., Li, X., Ogbazghi, A. Y., Feng, R., Dai, Z., Marchenkov, A., Conrad, E., First, P., and DeHeer, W. A.: Ultrathin epitaxial graphite: 2d electron gas properties and a route toward graphene-based nanoelectronics, *J. Phys. Chem. B* vol. 108, no. 52, pp. 19912–19916, 2004.

[5] Xu, C., Li, H., and Banerjee, K.: Modeling, analysis, and design of graphene nanoribbon interconnects, *IEEE Trans. Electron Devices* vol. 56, no. 8, pp. 1567–1578, 2009.

[6] Murali, R., Brenner, K., Yang, Y., Beck, T., and Meindl, J. D.: Resistivity of graphene nanoribbon interconnects, *IEEE Electron Device Lett.* vol. 30, no. 6, pp. 611–613, 2009.

[7] Cui, J. P., Zhao, W. S., Yin, W. Y., and Hu, J.: Signal transmission analysis of multilayer graphene nano-ribbon (MLGNR) interconnects, *IEEE Trans. Electromagn. Compat.* vol. 54, no. 1, pp. 126–132, 2012.

[8] Zhao, W. S., and Yin, W. Y.: Comparative study on multilayer graphene nanoribbon (MLGNR) interconnects, *IEEE Trans. Electromagn. Compat.* vol. 56, no. 3, pp. 638–645, 2014.

[9] Naeemi, A., and Meindl, J. D.: Compact physics-based circuit models for graphene nanoribbon interconnects, *IEEE Trans. Electron Devices* vol. 56, pp. 1822–1833, 2009.

[10] Naeemi, A., and Meindl, J. D.: Conductance modeling for graphene nanoribbon (GNR) interconnects, *IEEE Electron Device Lett.* vol. 28, pp. 428–431, 2007.

[11] Sarkar, D., Xu, C., Li. H., and Banerjee, K.: High-frequency behavior of graphene-based interconnects-Part I: Impedance modeling, *IEEE Trans. Electron Devices* vol. 58, pp. 843–852, 2011.

[12] Kumar, V., Rakheja, S., and Naeemi, A.: Performance and energy-per-bit modeling of multilayer graphene nanoribbon conductors, *IEEE Trans. Electron Devices* vol. 59, pp. 2753–2761, 2012.

[13] Kumar, V., Rakheja, S., and Naeemi, A.: Modeling and optimization for multilayer graphene nanoribbon conductors. In: *2011 IEEE International Interconnect Technology Conference*, Dresden, Germany, 2011, pp. 1–3, doi: 10.1109/IITC.2011.5940340.

[14] Bhattacharya, S., Das, S., Mukhopadhyay, A., Das, D., and Rahaman, H.: Analysis of temperature dependent delay optimization model for GNR interconnect using wire sizing method, *J. Comput. Electron.* vol. 17, no. 4, pp. 1536–1548, 2018.

[15] Das, S., Bhattacharya, S., Das, D., and Rahaman, H.: Thermal stability analysis of graphene nano-ribbon interconnect and applicability for terahertz frequency, *Natl. Acad. Sci. Lett.* vol. 43, no. 5, pp. 253–257, 2019.

[16] Bhattacharya, S., Das, D., and Rahaman, H.: Stability analysis in top contact and side contact graphene nanoribbon interconnects, *IETE J. Res.* vol. 63, no. 4, pp. 588–596, 2017.

[17] ITRS International Technology Working Groups: *International Technology Roadmap for Semiconductors*, ITRA, 2012.

[18] Chiariello, A. G., Maffucci, A., and Miano, G.: Temperature effects on electrical performance of carbon-based nano-interconnects at chip and package level, *Int. J. Num. Model.* vol. 26, pp. 560–572, 2013.

[19] Im, S., Srivastava, N., Banerjee, K., and Goodson, K. E.: Scaling analysis of multilevel interconnect temperatures for high-performance ICs, *IEEE Trans. Electron Devices* vol. 52, p. 2710, 2005.

[20] Maffucci, A., Micciulla, F., Cataldo, A., Bellucci, S., and Miano, G.: Electrothermal modeling and characterization of carbon interconnects with negative temperature coefficient of the resistance. In: *Proceedings of the 20th IEEE Workshop Signal Power Integr*, SPI, Turin, Italy, May 2016, pp. 9–11.

[21] Das, S., Das, D., and Rahaman, H.: RF performance analysis of graphene nanoribbin interconnect. In: *IEEE TechSym 2014–2014 IEEE Students' Technology Symposium*, pp. 105–110. doi: 10.1109/TechSym.2014.6807923.

[22] Jiang, J., Kang, J., and Banerjee, K.: Characterization of self-heating and current-carrying capacity of intercalation doped graphene-nanoribbon interconnects. In: *Proceedings of Reliability Physics Symposium (IRPS), 2017 IEEE International* Monterey, April 4–6, 2017, pp. 6B.1.1–6B.1.6.

[23] Rai, M. K., Arora, S., and Kaushik, B. K.: Temperature-dependent modeling and performance analysis of coupled MLGNR interconnects, *Int. J. Circ. Theory Appl.* vol. 46, no. 2, pp. 299–312, 2018.

[24] Bhattacharya, S., Das, D., and Rahaman, H.: Analysis of temperature dependent power supply voltage drop in graphene nanoribbon and Cu based power interconnects, *AIMS Mater. Sci.* vol. 3, no. 4, pp. 1493–1506, 2016.

[25] Das, S., Das, D., and Rahaman, H.: Electro-thermal RF modeling and performance analysis of graphene nanoribbon interconnects, *J. Comput. Electron.* vol. 17, pp. 1695–1708, 2018.

[26] Tunga, S. G., Bhattacharya, S., Das, S., and Rahaman, H.: Modeling of Pristine and intercalation doped multilayer graphene nanoribbon conductors with energy-per-layer screening. In: *Proceedings of the 5th International Symposium on Devices, Circuits and Systems (ISDCS)*, Springer, New York, 2022.

[27] Nishad, A., and Sharma, R.: Analytical time-domain models for performance optimization of multilayer GNR interconnects, *IEEE J. Sel. Top. Quant. Electron.* vol. 20, no. 1, pp. 17–24, 2014.

[28] Das, S., Bhattacharya, S., and Das, D.: Analysis of stability in carbon nanotube and graphene nanoribbon interconnects, *Int. J. Soft Comput. Eng.* vol. 2, no. 6, pp. 325–329, January 2013. ISSN: 2231-2307.

[29] Bhattacharya, S., Das, D., and Rahaman, H.: Reduced thickness interconnect model using GNR to avoid crosstalk effects, *J. Comput. Electron.* vol. 15, pp. 367–380, 2016.

[30] Stellari, F., and Lacatia, A. L.: New formulas of interconnect capacitances based on results of conformal mapping method, *IEEE Trans. Electron Devices* vol. 47, no. 1, pp. 222–231, 2000.

[31] Nishad, A. K., and Sharma, R.: Self-consistent capacitance model for multilayer graphene nanoribbon interconnects, *Micro Nano Lett.* vol. 10, no. 8, pp. 404–407, 2015.

[32] Gantmakher, V. F.: The experimental study of electron -phonon scattering in metals, *Rep. Prog. Phys.* vol. 37, pp. 317–361, 1974.

[33] Banerjee, K., and Mehrotra, A.: Analysis of on-chip inductance effects for distributed RLC interconnects, *IEEE Trans. Comput. Aided Des. Integr. Circ. Syst.* vol. 21, no. 8, pp. 904–915, 2002.

[34] Kumar, M. G., Chandel, R., and Agrawal, Y.: Timing and stability analysis of carbon nanotube interconnects. In: *IEEE International Symposium on Nanoelectronics and Information Systems* (2015), Indore, MP, 21–23 Dec. 2015, pp. 308–313. doi: 10.1109/iNIS.2015.43.

6 Electro-Thermal Modeling of CNT and GNR Interconnect for Nano-Electronic Circuits

Santasri Giri Tunga, Subhajit Das,
and Hafizur Rahaman

6.1 INTRODUCTION

Graphene has inspired the integrated chip industry researchers to prove it as a prospective interconnect material in the nanoscale range. Graphene can be rolled up to form a carbon nanotube (CNT) or cut into flat narrow strips to form graphene nanoribbon (GNR) as depicted in Figure 6.1 whose electrical properties are determined by the edge chirality and width [1]. Quite a lot of research developments have been analysed with CNT and GNR, which substantially evinced that in terms of conductivity (both electrical and thermal), mean free path (MFP), parasitic and skin effects, stress, etc., these two materials supersede copper (Cu) as the future interconnect material [2–4].

CNT and GNR exhibit both metallic and semiconducting property [5]. Categorized by edge geometry, CNT and GNR can be armchair or zigzag [6]. The major difference between CNT and GNR lies in the fact that, armchair CNT is metallic in nature while zigzag CNT can be both metallic and semiconducting, while for GNR, metallic property is exhibited by zigzag GNR and armchair GNR shows both metallic and semiconducting property [7]. The metallic CNT and GNR find application

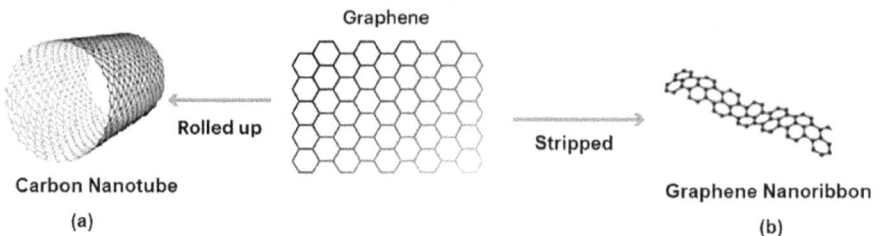

Graphene

Rolled up

Carbon Nanotube

(a)

Stripped

Graphene Nanoribbon

(b)

FIGURE 6.1 Structures of graphene interconnects: (a) CNT and (b) GNR.

DOI: 10.1201/9781003331650-6

in nano-interconnects while semiconducting property is widely used for nano-electronic devices.

A single layer of graphene when rolled up with a specific diameter (*Dia*) forms Single-Walled CNT (SWCNT). Concentric arrangement of many SWCNT gives rise to metallic natured Multi-Walled CNT (MWCNT) [8]. The number of SWCNT in a MWCNT is diameter dependent as shown in Figure 6.2, and the diameters of each rolled layer differ from one another. The individual rolled layer is known as shell. Adjacent shells maintain a minimum distance of 0.34 nm in between them known as van der walls gap. The diameter of the innermost shell is smallest and outermost shell is largest among all. If the largest diameter is Dia_{max} and smallest diameter is Dia_{min}, then the number of shells (N_{CNT_shell}) found in a MWCNT is calculated as [9,10]

$$N_{CNT\_shell} = k = 1 + int\left[\frac{Dia_{max} - Dia_{min}}{2\partial_s}\right]$$ (6.1)

If the shells are numbered from the outer to innermost direction as 1, 2, 3,..., j,..., k then jth shell and innermost shell will have a diameter equal to [10]

$$Dia_j = Dia_{max} - 2\partial_s(k-1) \; 1 \le j \le k$$ (6.2)

$$Dia_{min} = Dia_{max} - 2\partial_s(k-1)$$ (6.3)

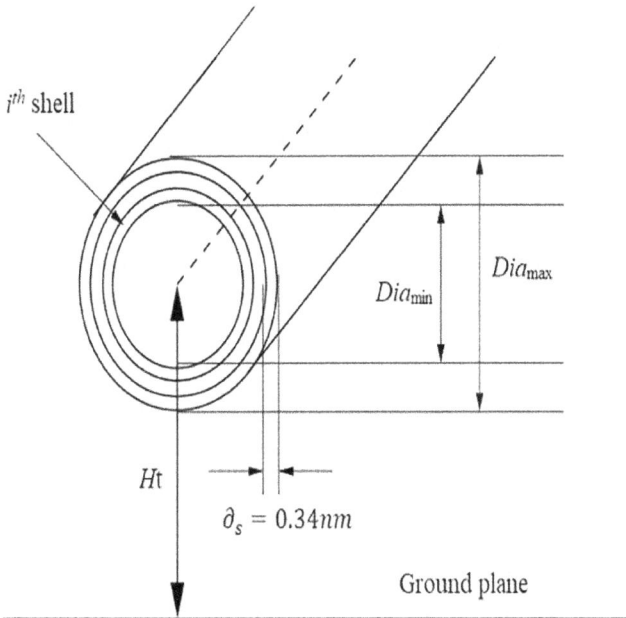

FIGURE 6.2   Multi-walled carbon nanotube over a ground plane.

Each shell has many conducting channels and in jth shell, the number of conducting channels is

$$N_{ch\_j} = aDia_j + b \qquad (6.4)$$

where a and b are 0.0612 per nm and 0.425, respectively [9].

The fundamental equivalent single conductor (ESC) model of MWCNT is depicted in Figure 6.3a and b. Length-independent contact resistance (R_{ct}) and quantum resistance (R_{qt}) constitute the lumped elements. Variation in conductance is found in the distributed elements which comprise the effective resistance (R_{efct}), effective inductance (L_{efct}), and effective capacitance (C_{efct}) [8]. Presuming a perfect metal-CNT contact, the quantum resistance equals 6.45 kΩ and imperfect contact resistance varies from 0 to 100 kΩ [8–10]. The R_{efct} is a function of length and temperature-dependent MFP. These are discussed in detail.

A monolayer of graphene sheet when stripped to very slender ribbons with a width less than 50 nm is termed as a single-layer GNR (SLGNR). The single layer can be suspended or placed on a substrate. However, in both states, SLGNR concedes high resistance and is ill-fitted for interconnect application [11,12]. To apt GNR for interconnect application, a number of SLGNRs are piled upon one another keeping a distance of 0.34 nm in between the layers, termed as multi-layer GNR (MLGNR) whereby the resistance is alleviated [13]. Different studies have revealed and explained the performance superiority of MLGNR over SLGNR [14,15]. Metal-GNR contact is made when MLGNR is placed in the interconnect system and the metal can be placed at the topmost layer (TC-MLGNR) or at the side connecting all layers (SC-MLGNR).

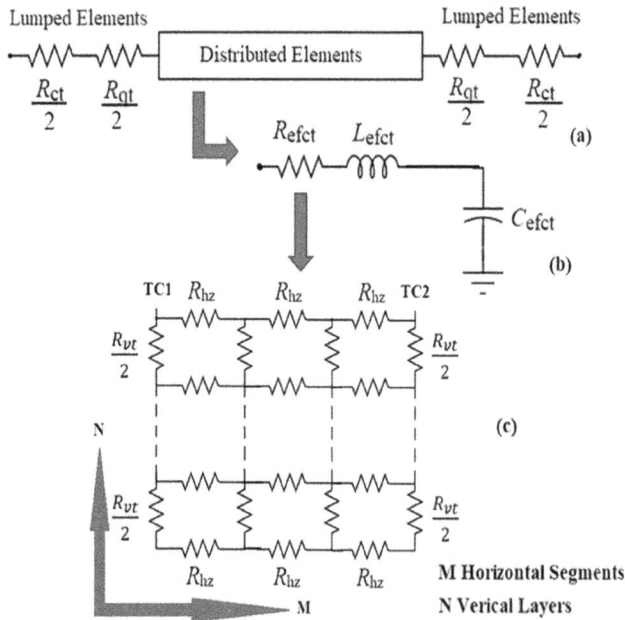

FIGURE 6.3   Fundamental ESC model of MLGNR.

Compared to SC-MLGNR, TC-MLGNR shows unsatisfactory and inferior performance [16,17]. However, fabrication complexity of SC-MLGNR in planar technology has switched researchers to improve performance of TC-MLGNR and make it suitable for interconnect application. Pronounced developments have been noticed when atoms of other elements are introduced in between the MLGNR layers called as intercalation doping. It has been observed and demonstrated that after intercalation doping, the Fermi level became high, the interlayer distance enlarged, the MFP enhanced, and subsequently, performance of TC-MLGNR approached SC-MLGNR [18–21].

The elementary equivalent single conductor (ESC) model of the MLGNR interconnect is depicted in Figure 6.3a–c. It comprises of lumped elements and distributed elements. Alteration in interconnect length does not affect the lumped elements' performance, whereas pronounced effect is detected in distributed elements. Similar to MWCNT, for MLGNR the quantum resistance is not constant but depends on the number of conducting channels calculated from the Fermi-Dirac equation and equals 12.94 kΩ per channel [11].

For MLGNR the effective resistance is the combination of the in-plane resistance along the horizontal segments and perpendicular resistance along the vertical layers stated mathematically as [15]

$$R_{hz} = \frac{R_{qt}\Delta len}{N_{ch}\lambda_{efct}} \tag{6.5}$$

$$R_{vt} = \frac{\sigma_c \partial_i}{W_d \Delta len} \tag{6.6}$$

The in-plane resistance (R_{hz}) is MFP dependent, which changes with variation in temperature and vertical resistance (R_{vt}) depends on the interlayer distance and c-axis resistivity. Both R_{hz} and R_{vt} are found to vary with length.

6.2 ELECTRO-THERMAL MODEL OF MWCNT AND MLGNR

Metallic MWCNT and MLGNR are used as interconnect material. Temperature has a direct impact upon the conductivity of metals, subsequently the performance of both MWCNT and MLGNR is found to vary with temperature [22]. The temperature-dependent MFP regulates the effective resistance of MLGNR and ohmic resistance in MWCNT.

MFP is determined by different scattering mechanism and the scattering can be controlled by temperature [23]. The mechanism of scattering as has been investigated can be (1) electron-electron scattering and (2) electron-phonon scattering caused by lattice vibration [14]. Electron-electron scattering is found at all temperatures, and prevails mostly at lower temperatures. Whenever the temperature goes higher, lattice vibration occurs, subsequently electron-phonon scattering plays a dominant and significant criterion to control the MFP [24]. As temperature has a dominant contribution on the electron-phonon scattering mechanism, in our present study we have disregarded electron-electron scattering and emphasized on the electron-phonon scattering for our model.

TABLE 6.1

Temperature Coefficient of Cu, CNT, and GNR [27]

Material	DC Conductivity	Temp. Coefficient (α)
Cu	5.96×10^7	0.003862
CNT	10^6–10^7	−0.000387
Graphene	10^8	−0.0002

With increasing temperature, lattice vibration increases, subsequently more energy is transferred due to the adsorption and radiation of phonon-electron interaction [24]. The aspect of this interaction is acoustic phonon scattering and optical phonon scattering. In MWCNT, electron-phonon scattering predominates and comprises of temperature- and diameter-dependent acoustic phonon scattering [25] and optical absorption and emission scattering [26]. Similarly, in MLGNR, it is found that the leading source of scattering is (1) edge scattering (λ_{edge}), (2) acoustic phonon scattering (λ_{ACP}), (3) optical phonon scattering consisting of optical emission and absorption (λ_{OP-abs}, λ_{OP-emn}), and (4) remote interfacial phonon scattering (λ_{RIP}). Apart from λ_{edge}, in MLGNR all other scatterings are temperature dependent. With increase in temperature, electrons get energized and start oscillating and collide more with other electrons, i.e., scattering of electrons increases. As a result, the mean free path shortens and resistance of the interconnect is augmented. Thus, with increased temperature, MFP reduces, subsequently the resistance increases.

As shown in Table 6.1, the conductivity from highest to lowest is found for graphene, CNT, and then Cu [27]. Besides, graphene and CNT have negative temperature coefficient, which otherwise reveals the fact that the heat generated with temperature increase will be dissipated more in CNT and graphene than that of Cu. Subsequently, CNT and graphene can be used in interconnect applications. Skin effect is also not prominent as these are only an atom in thickness.

6.3 MODELING OF TEMPERATURE-DEPENDENT MWCNT CONDUCTORS

For MWCNT the total resistance is the sum of the contact resistance (R_{ct}), the quantum resistance (R_{qt}), and the ohmic resistance (R_{oh}) written as

$$Rt_{CNT} = R_{ct} + R_{qt} + R_{oh} \tag{6.7}$$

$$R_{oh} = \frac{h \, Len_{CNT}}{4e^2 \lambda_{efct}(T)} \tag{6.8}$$

From equation (6.8) it can be said that the ohmic resistance varies with temperature-dependent MFP. MFP is determined by different scattering mechanisms and the scattering can be controlled by temperature (Temp). The mechanism of scattering can be electron-electron or electron-phonon. In MWCNT, electron-phonon scattering predominates and comprises of temperature-dependent acoustic phonon scattering and optical absorption and emission scattering.

Besides temperature, diameter is also a factor upon which the MFP of MWCNT is dependent. So, diameter and temperature-dependent acoustic phonon scattering MFP is determined as [26]

$$\lambda_{AcPh} = 400 \times 10^3 \times \frac{Dia}{Temp} \tag{6.9}$$

Apart from acoustic phonon scattering, the MFP due to optical phonon scattering also prevails in MWCNT. It consists of optical phonon absorption and optical phonon emission. Two significant factors that determine the optical phonon absorption and emission are the phonon occupation number and the spontaneous optical phonon emission. The impetuous scattering of optical phonon is diameter dependent and expressed as $\lambda_{OPh} = 56 \times Dia$, whereas the threshold energy of optical phonon emission ($\hbar\omega_{OPh} \approx 0.16 - 0.2\,eV$) and the Boltzmann constant determine the phonon emission and absorption occupation number expressed as

$$N_{OPh} = \left(e^{\frac{\hbar\omega_{OP}}{KbT}} - 1 \right)^{-1} \tag{6.10}$$

Optical phonon absorption at any temperature is determined as

$$\lambda_{OPhAbs}(Temp) = \lambda_{OPh}\left(\frac{N_{OPh}(300)+1}{N_{OPh}(Temp)} \right) \tag{6.11}$$

However, optical phonon emission is governed by the cumulative effect of absorption of optical phonon ($\lambda_{OP_{emn}}^{Abs}$) and the applied field ($\lambda_{OP_{emn}}^{F}$). Both these factors are temperature dependent and determined as [28]

$$\lambda_{OPh_{emn}}^{F}(Temp) = \frac{\hbar\omega_{OPh} - KbT}{eV/Len_{CNT}} + \lambda_{OPh}\left(\frac{N_{OPh}(300)+1}{N_{OPh}(Temp)+1} \right) \tag{6.12}$$

$$\lambda_{OPh_{emn}}^{Abs}(Temp) = \lambda_{OPhAbs} + \lambda_{OPh}\left(\frac{N_{OPh}(300)+1}{N_{OPh}(Temp)+1} \right) \tag{6.13}$$

Therefore, the total optical phonon emission at any temperature is written as

$$\lambda_{OPhemn}(Temp) = \lambda_{OPh_{emn}}^{Abs}(Temp) + \lambda_{OPh_{emn}}^{F}(Temp) \tag{6.14}$$

The cumulation of all the above scatterings leads to the effective MFP of MWCNT resolved as per Matthiessen's rule

$$\frac{1}{\lambda_{efct}(Temp)} = \frac{1}{\lambda_{AcPh}(Temp)} + \frac{1}{\lambda_{OPh_{abs}}(Temp)} + \frac{1}{\lambda_{OPhemn}(Temp)} \tag{6.15}$$

Figure 6.4 demonstrates that MFP of MWCNT is directly proportional to diameter of CNT and inversely proportional to temperature. Reduction of the MFP at higher

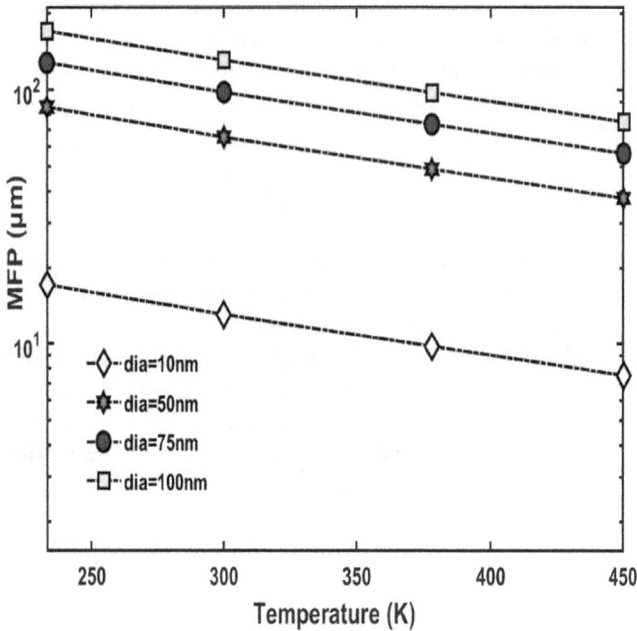

FIGURE 6.4 Variation of MFP of MWCNT with temperature for different diameters.

temperature causes more scattering, subsequently increasing the resistance. So, the total resistance of MWCNT as is observed in Figure 6.5 increases with increase in temperature and length.

Apart from resistors, the distributed elements comprise of the inductors and capacitors. The inductance is a combination of magnetic inductance and kinetic inductance. Magnetic inductance has been ignored because of its insignificant value and diameter-dependent kinetic inductance becomes notable. The kinetic inductance of MWCNT is determined as [25]

$$L_{kn\_MWCNT} = \left(\sum_{j=1}^{k} \frac{1}{L_{kn}^{j}} \right)^{-1} \tag{6.16}$$

where L_{kn}^{j} is the jth shell kinetic inductance and equals $16/2\,N_{ch\_j}$ nH/μm.

For MWCNT, the quantum capacitance is determined as [25]

$$C_{qt\_MWCNT} = \sum_{j=1}^{k} C_{qt}^{j} \tag{6.17}$$

where C_{qt}^{j} is the jth shell quantum capacitance and equals $2 \times 96.8\,N_{ch\_j}$ aF/μm. Besides, MWCNT experiences electrostatic capacitance between the adjacent shells and also between the outermost diameter and the ground. However, unlike resistance, the inductance and capacitance of MWCNT are indifferent to temperature variation.

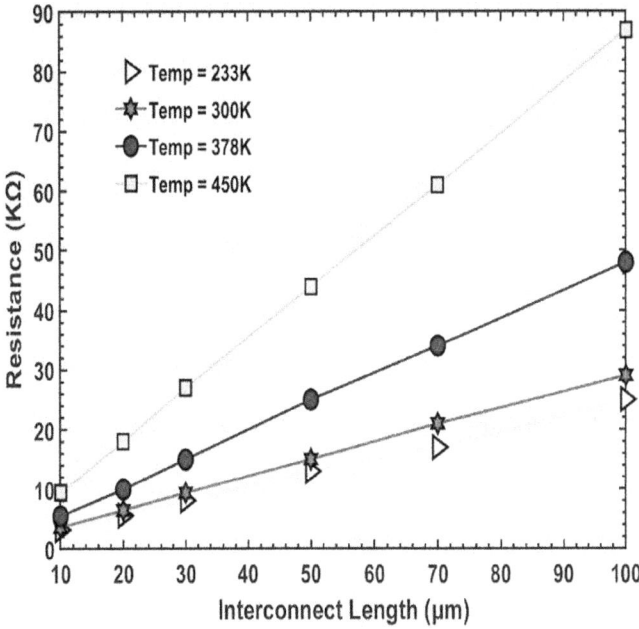

FIGURE 6.5 Variation of resistance of MWCNT with interconnect length for different temperatures.

6.4 MODELING OF TEMPERATURE-DEPENDENT MLGNR CONDUCTORS

Equations (6.5) and (6.6) give a relation between the resistance with interconnect length, width, number of conducting channels, the c-axis resistivity and finally the temperature-dependent MFP [29]. The Fermi level dependent N_{ch} in MLGNR changes nominally with temperature as formulated by the Fermi-Dirac equation [23]:

$$N_{ch} = \sum_{b=1}^{nc} \left(1 + e^{\frac{E_f + E_b}{K_b T}} \right)^{-1} + \sum_{b=1}^{nv} \left(1 + e^{\frac{E_b - E_f}{K_b T}} \right)^{-1} \tag{6.18}$$

E_b is defined as the energy of the highest subband for valance band and lowest subband for conduction band and expressed as [11]

$$E_b = \frac{h v_f}{2 W_m} |b + \beta| \tag{6.19}$$

As can be seen in Figure 6.6, N_{ch} increases nominally for a large variation in temperature for pristine as well as intercalation-doped MLGNR [23]. However, N_{ch} for different widths is found to vary by a significant value for pristine and intercalation-doped MLGNR as the subband energy is greatly influenced by width of the interconnect as demonstrated in equation (6.18).

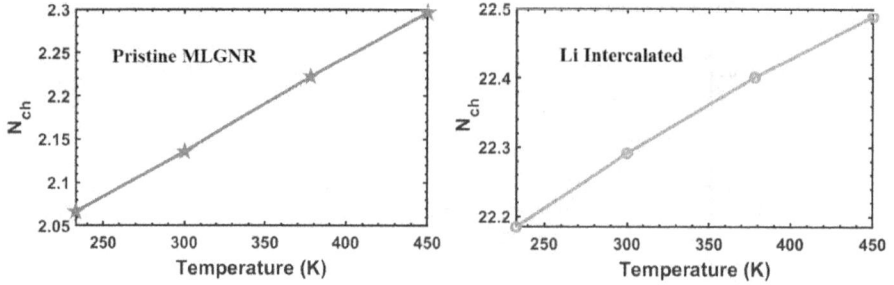

FIGURE 6.6 Variation of number of conducting channels with temperature for pristine and Li intercalation-doped MLGNR.

MFP-dependent scattering resistance has been considered as a prime criterion for modeling the electro-thermal GNR interconnect determined as

$$R_s = \frac{R_{qt}\Delta len}{\lambda_{efct}} \tag{6.20}$$

MFP in GNR, similar to MWCNT, is limited by the scattering of electrons and phonons [30] and is determined by different scattering mechanisms and the scattering can be controlled by temperature (Temp). Lattice vibration at higher temperature results in emission of electrons due to disturbance of potential. At this point the scattered electrons start to transfer energy with local phonons and oscillate with them either in-phase or out-of-phase. In-phase oscillation of electron-phonon is known as acoustic phonon scattering, and out-of-phase oscillation of electron-phonon is termed as optical phonon scattering [31].

The MFP arising due to acoustic phonon scattering is determined as [14]

$$\lambda_{ACP}(\text{Temp}) = 4 \frac{\rho_m \left(\hbar v_f v_s\right)^2}{D_{ACP}^2 \sqrt{\pi N_{C2D}} K_b T} \tag{6.21}$$

where D_{ACP}, the acoustic deformation potential, varies in the range of 6–30 eV, the 2D mass density is ρ_m, acoustic phonon velocity is v_s, and N_{C2D} is the concentration of 2DEG in graphene. The above-mentioned parameters are constant and possess very high value, eventually, the change in temperature has an insignificant effect on λ_{ACP} as can be observed in Figure 6.7.

Optical phonon scattering comprises of both phonon emission and phonon absorption [14,23] and the MFP generated by these two types of scattering is represented as

$$\lambda_{OPabs-emn}(\text{Temp}) = \lambda_{OP-abs}(\text{Temp}) + \lambda_{OP-emn}(\text{Temp}) \tag{6.22}$$

where

$$\lambda_{OP-abs}(\text{Temp}) = \frac{\rho_m \hbar \omega_{OP} v_f^2}{D_{op}^2 \sqrt{\pi N_{C2D}} N_{OP-abs}\left(1 + \dfrac{\omega_{OP}}{v_f \sqrt{\pi N_{C2D}}}\right)} \tag{6.23}$$

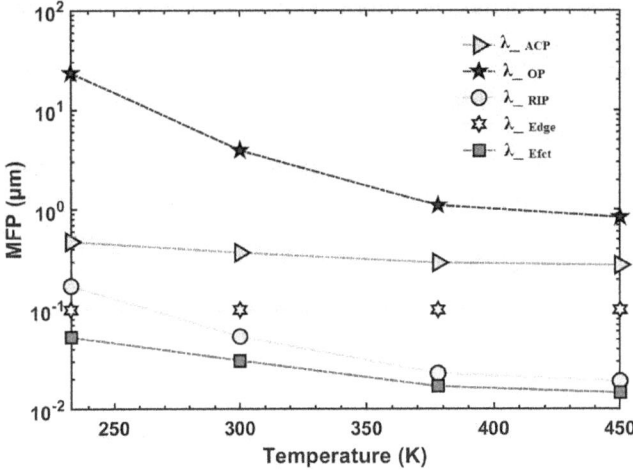

FIGURE 6.7   Temperature vs different types of MFP of MLGNR.

$$\lambda_{OP-emn}\left(\text{Temp}\right) = \frac{\rho_m \hbar \omega_{OP} v_f^2}{D_{op}^2 \sqrt{\pi N_{C2D}} N_{OP-emn}\left(1 - \dfrac{\omega_{OP}}{v_f \sqrt{\pi N_{C2D}}}\right)} \qquad (6.24)$$

where $\hbar \omega_{OP}$ is optical phonon energy (≈ 160 meV), D_{op} is optical deformation potential ranging between 100×10^9 and 400×10^9 eV/m. N_{OP}, the phonon occupation number, is constant both for optical emission and absorption and equals $1/(e^{\hbar \omega_{OP}/K_b T} - 1)$. It can be observed from Figure 6.7 that with gradual rise in temperature the optical phonon scattering indicates an exponential sharp decrease, which proves a strong impact of temperature upon it.

The mechanism in which the optical phonons at the surface of the substrate remotely combine with the electrons in GNR layer gives rise to remote interfacial phonon (RIP) scattering [32]. The potency of the RIP scattering (β_{RIP}) is a function of the dielectric constant of the substrate upon which the GNR layer is deposited and is determined as [33]

$$\beta_{RIP} = \frac{\epsilon_{st} - \epsilon_{\omega}}{(\epsilon_{st} + 1)(\epsilon_{\omega} + 1)} \qquad (6.25)$$

where ϵ_{ω} and ϵ_{st} are the high-frequency dielectric constant and static dielectric constant of the dielectric substrate.

The MFP for RIP scattering is expressed as [21]

$$\lambda_{RIP}\left(\text{Temp}\right) = \beta_{RIP}\ W_m \left(\frac{E_f}{q}\right)^{\alpha}\left(e^{\frac{E_0}{K_b T}} - 1\right) \qquad (6.26)$$

E_0 and α are considered as two global parameters with values equal to 104 and 1.02 meV, respectively [34,35]. Similar to optical phonon scattering, remote interfacial phonon scattering also decreases exponentially with increase in temperature as demonstrated in Figure 6.7.

Besides scattering of electrons and phonons, edge scattering is an important factor in MLGNR upon which MFP depends. In view of edge roughness, if the edge is very smooth, termed as fully specular edge ($p = 1$), there is least or no alteration in MFP [11]. However, fully specular edge is almost unachievable. So, GNRs fabricated as yet have mostly diffusive edge where the specularity constant (p) varies from 0.2 to 0.8. The diffusive scattering MFP at the edges is a contribution of the effective subbands of each layer, the Fermi energy, Fermi velocity, and width of GNR and is determined from

$$\lambda_{edge} = \frac{W_m}{1-p} \sqrt{\left(\frac{2W_m Ef}{jhv_f}\right)^2 - 1} \qquad (6.27)$$

where p is chosen as zero and unity for fully diffusive and fully specular edge GNR, respectively. The variation of MFP due to edge scattering is indifferent to temperature, subsequently, it remains constant at all temperatures.

Hence, the MFP of MLGNR considering all the scattering leads to the effective MFP determined by Matthiessen's rule

$$\left(\lambda_{efct}\right)^{-1} = \frac{1}{\left(\lambda_{edge}^{-1} + \lambda_{ACP}^{-1} + \lambda_{OP_{abs-emn}}^{-1} + \lambda_{RIP}^{-1}\right)^{-1}} \qquad (6.28)$$

Temperature-dependent MFP determines the resistance, and with increase in temperature MFP reduces. Consequently, resistance of MLGNR increases with temperature, thereby reducing conductivity. As mentioned earlier, to boost the conductivity of MLGNR, intercalation doping is an effective way [23]. For intercalation-doped MLGNR, MFP is more than pristine MLGNR and hence the conductivity is enhanced.

From Figure 6.8, it can be observed that compared to pristine MLGNR, different intercalation-doped MLGNR has higher MFP. Intercalation doping increases the charge carriers which is responsible for upward shift of the Fermi level [21] and eventually broadens the interlayer distance. Among AsF_5, $FeCl_3$, Li, and $MoCl_5$ intercalation doping, the latter has the highest Fermi level and larger interlayer distance that abates the scattering phenomenon. In Table 6.2 the properties of pristine and different intercalation-doped MLGNR are listed.

The resistance modeling of pristine MLGNR (SC and TC) and different intercalation-doped TC-MLGNR is determined from [37] where the cumulation of the horizontal resistance of M segments and vertical resistances of N layers gives the effective resistance. The horizontal resistance as given in equation (6.5) is a function of temperature-dependent N_{ch} and MFP. In TC-MLGNR, as the top most layer has direct contact with metal, the effect of perpendicular or vertical resistance needs to be considered; however, in SC-MLGNR all layers are in contact with metal, so there is absence of vertical resistance depicted in Figure 6.9.

$$R_{efct\_TC-MLGNR} = \left(\frac{1}{R_{hz(N-1)} + R_{vt(N)}} + \frac{1}{R_{hz(N)}}\right)^{-1} \qquad (6.29)$$

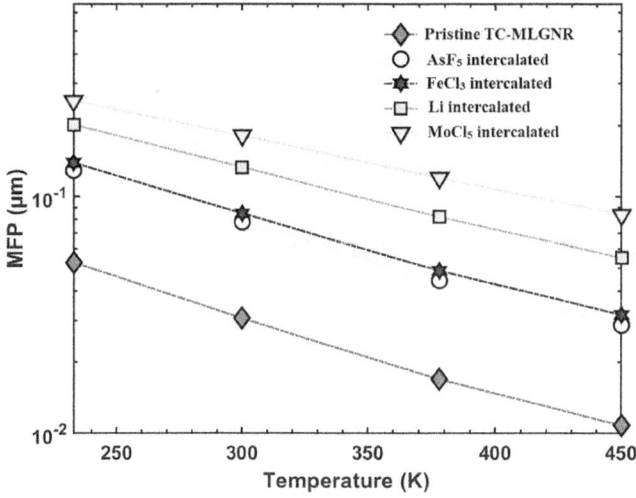

FIGURE 6.8 Variation of MFP with temperature for pristine and different intercalation-doped MLGNR.

TABLE 6.2
Properties of Pristine- and Intercalation-Doped MLGNR [36]

Properties	Pristine MLGNR TC	AsF$_5$ Intercalated TC	FeCl$_3$ Intercalated TC	Li Intercalated TC	MoCl$_5$ Intercalated TC
Fermi energy (eV)	0.2	0.6	0.68	1.5	3.4
Van der Walls gap (nm)	0.34	0.575	0.394	0.37	0.92
Number of conducting channel	2	8	10	22	50
C-axis conductivity (S/cm)	0.033	0.24	1	1.8×10^4	0.33

$$R_{efct\_SC-MLGNR} = \left(\frac{R_{hz}}{N} \right) \tag{6.30}$$

As can be determined from equations (6.29) and (6.30), the effective resistance of SC-MLGNR is very less compared to TC-MLGNR and different intercalation-doped TC-MLGNRs for a wide temperature range.

In the ESC model of Figure 6.3, it can be seen that R_{efct} is a part of the distributed element which combined with the lumped resistances to give the total resistance. Hence the total resistance per unit length of TC-MLGNR is the aggregate of the lump resistance consisting of the contact and quantum resistance and the distributed or effective resistance consisting of the horizontal and perpendicular resistance given as [37,38]

$$R_{Total\_TC} = R_{efct\_TC-MLGNR} + \frac{R_{qt} + R_{Tct1} + R_{Tct2}}{N_{ch}} \tag{6.31}$$

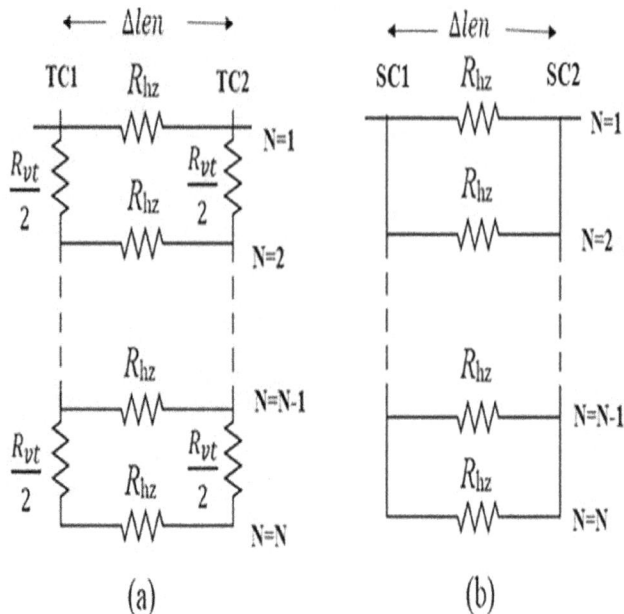

FIGURE 6.9 Per unit length resistance for N layers of (a) top contact and (b) side contact MLGNR [35].

Contact resistance for TC-MLGNR has been calculated as 4.3 kΩ.

Likewise, the total resistance per unit length of the SC-MLGNR is the aggregate of the lump resistance consisting of the contact and quantum resistance and the distributed or effective resistance consisting of the horizontal resistance given as [37,38].

$$R_{Total\_SC} = R_{efct\_SC-MLGNR} + \frac{R_{qt} + R_{Sct1} + R_{Sct2}}{N_{ch} * N_{layer}} \qquad (6.32)$$

Contact resistance for SC-MLGNR has been calculated as 100 Ω.

In Figure 6.10, variation of total resistance with temperature for pristine (TC and SC) MLGNR and for different types of intercalation-doped TC-MLGNR is portrayed. The comparison of the total resistance with temperature for MLGNR, MWCNT with copper is illustrated in Figure 6.11. Along with resistance, the ESC model depicted in Figure 6.3b consists of effective inductance and capacitance. The cumulative effect of magnetic inductance (L_{Mg}) and kinetic inductance (L_{Kn}) results in the effective inductance (L_{efct}). The value of L_{Mg} has been discarded in MLGNR conductance modeling owing to its trivial value. Kinetic inductance is a function of number of layers and N_{ch} determined as [39]

$$L_{Kn} = \frac{h}{4e^2 V_f N_L N_{ch}} \qquad (6.33)$$

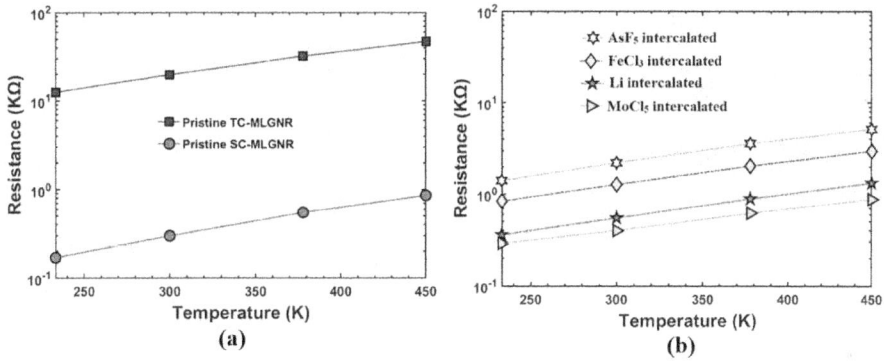

FIGURE 6.10 Variation of total resistance with temperature for (a) TC- and SC-MLGNR and (b) different types of intercalation-doped TC-MLGNR.

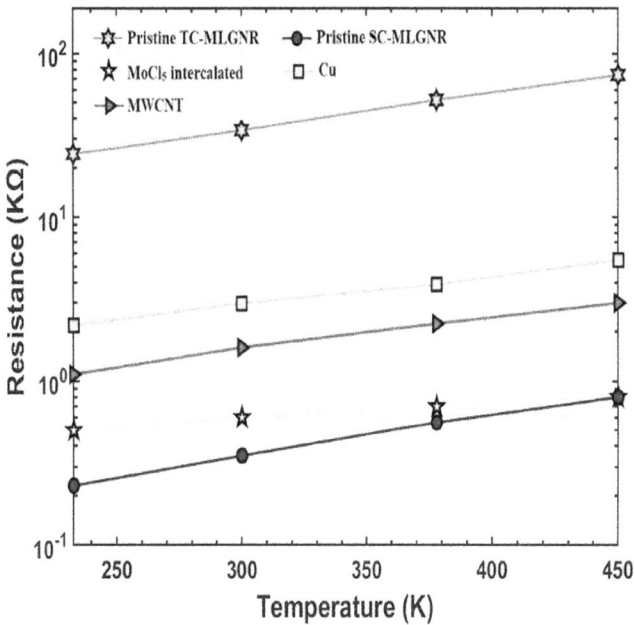

FIGURE 6.11 Variation of total resistance with temperature for Cu, MWCNT, Pristine and intercalation-doped TC-MLGNR and SC-MLGNR.

N_{ch} differs slightly with temperature, subsequently, variation in temperature does not affect much in the value of kinetic inductance. Likewise, deviation in effective capacitance with temperature is minimum and comprises of the quantum capacitance (C_{qc}) [37] and electrostatic capacitance (C_{ec}) [40,41] formulated as

$$C_{efct} = \left(\frac{1}{C_{qc}} + \frac{1}{C_{ec}} \right)^{-1} \qquad (6.34)$$

6.5 CONCLUSION

This work presents a comprehensive temperature-dependent model comprising of resistance, inductance, and capacitance of MWCNT and MLGNR interconnect. For different Fermi energy and temperature, a number of conducting channels have been found to differ. Further, considering temperature-driven variation in electron and phonon scattering mechanism, MFP has been computed and is incorporated in the model. The developed RLC model of the nano-interconnect in view of temperature variation reveals acceptable compliance with the experimental outcome which subsequently depicts temperature dependency of resistance of TC-, SC-MLGNR, and MWCNT-based interconnects. It is also demonstrated that intercalation-doped MLGNR interconnects achieve remarkable and superior electrical performance at higher chip operating temperatures.

REFERENCES

[1] Xu, C., Li, H., and Banerjee, K.: "Modeling, analysis, and design of graphene nano-ribbon interconnects," *IEEE Trans. Electron Devices* vol. 56, no. 8, pp. 1567–1578, 2009.

[2] Balandin, A., Ghosh, S., Bao, W., Calizo, I., Teweldebrhan, D., Miao, F., and Lau, C. N.: "Superior thermal conductivity of single-layer graphene," *Nano Lett.* vol. 8, no. 3, pp. 902–907, 2008.

[3] Bolotin, K. I., Sikes, K. J., Hone, J., Stormer, H. L., and Kim, P.: "Temperature-dependent transport in suspended graphene," *Phys. Rev. Lett.* vol. 101, no. 9, p. 096802, 2008

[4] Murali, R., Brenner, K., Yang, Y., Beck, T., and Meindl, J. D.: "Resistivity of graphene nanoribbon interconnects," *IEEE Electron Device Lett.* vol. 30, no. 6, pp. 611–613, 2009.

[5] ITRS International Technology Working Groups: *International Technology Roadmap for Semiconductors*, ITRS, 2013.

[6] Naeemi, A., and Meindl, J. D.: "Compact physics-based circuit models for graphene nanoribbon interconnects," *IEEE Trans. Electron Devic.* vol. 56, no. 9, pp. 1822–1833, 2009.

[7] Saito, R., Dresselhaus, G., and Dresselhaus, M. S.: *Physical Properties of Carbon Nanotubes*, first edition, Imperial College Press, London, 1998.

[8] Burke, P. J.: "Liquid theory as a model of the gigahertz electrical properties of carbon nanotubes," *IEEE Trans. Nanotechnol.* vol. 1, no. 3, pp. 119–144, 2002.

[9] Naeemi, A., and Meindl, J. D.: "Compact physical models for multiwall carbon-nanotube interconnects," *IEEE Electron Device Lett.* vol. 27, no. 5, pp. 338–340, 2006.

[10] Naeemi, A., and Meindl, J. D.: "Design and performance modeling for singlewalled carbon nanotubes as local, semiglobal, and global interconnects in gigascale integrated systems," *IEEE Trans. Electron Devices* vol. 54, no. 1, pp. 26–37, 2007.

[11] Naeemi, A., and Meindl, J. D.: "Conductance modeling for graphene nanoribbon (GNR) interconnects," *IEEE Electron. Device Lett.* vol. 28, no. 5, pp. 428–431, 2007.

[12] Murali, R., Yang, Y., Beck, T., and Meindl, J. D.: "Resistivity of graphene nanoribbon interconnects," *IEEE Electron. Device Lett.* vol. 30, no. 6, pp. 611–613, 2009.

[13] Berger, C., Song, Z., Li, X., Wu, X., Brown, N., Naud, C., Mayou, D., Li, T., Hass, J., Marchenkov, A. N., Conrad, E. H., First, P. N., and Heer, W. A.: "Electronic confinement and coherence in patterned epitaxial graphene," *Science* vol. 312, pp. 1191–1196, 2006.

[14] Rakheja, S., Kumar, V., and Naeemi, A.: "Evaluation of the potential performance of graphene nanoribbons as on-chip interconnects," *Proc. IEEE* vol. 101, no. 7, pp. 1740–1765, 2013.

[15] Kumar, V., Rakheja, S., and Naeemi, A.: "Performance and energy-per-bit modeling of multilayer graphene nanoribbon conductors," *IEEE Trans. Electron Devices* vol. 59, pp. 2753–2761, 2012.

[16] Kumar, V., Rakheja, S., and Naeemi, A.: "Modeling and optimization for multilayer graphene nanoribbon conductors," *IEEE International Interconnect Technology Conference and 2011 Materials for Advanced Metallization*, IITC/MAM, 2011. doi: 10.1109/IITC.2011.5940340

[17] Bhattacharya, S., Das, D., and Rahaman, H.: "Stability analysis in top contact and side contact graphene nanoribbon interconnects," *IETE J. Res.* vol. 63, no. 4, pp. 588–596, 2017.

[18] Das, S., Bhattacharya, S., Das, D., and Rahaman, H.: "Comparative stability analysis of pristine and AsF_5 intercalation doped top contact graphene nano ribbon interconnects," *2019 2nd International Symposium on Devices, Circuits and Systems (ISDCS)*, Higashi-Hiroshima, Japan, 2019, pp. 1–4. doi: 10.1109/ISDCS.2019.8719094

[19] Tunga, S. G., Das, S., and Rahaman, H.: "A brief review of recent studies on performance improvement of graphene nanoribbon interconnect," *2021 International Symposium on Devices, Circuits and Systems (ISDCS)*, Higashihiroshima, Japan, 2021, pp. 1–6. doi: 10.1109/ISDCS52006.2021.9397920.

[20] Bao, W., Wan, J., Han, X., Cai, X., Zhu, H., Kim, D., Ma, D., Xu, Y., Munday, J. N., Dennis Drew, H., Fuhrer, and M. S., Hu, L.: "Approaching the limits of transparency and conductivity in graphitic materials through lithium intercalation," *Nat. Commun.* vol. 5, p. 4224, 2014.

[21] Jiang, J., Kang, J., Cao, W., Xie, X., Zhang, H., Chu, J., Liu, W., Banerjee, K.: "Intercalation doped multilayer-graphene nanoribbons for next-generation interconnects," *Nano Lett.* vol. 17, no. 3, pp. 1482–1488, 2017.

[22] Pop, E., Mann, D. A., Goodson, K. E., and Dai, H.: "Electrical and thermal transport in metallic single-wall carbon nanotubes on insulating substrates," *J. Appl. Phys.* vol. 101, no. 9, p. 093710, 2007.

[23] Das, S., Das, D., and Rahaman, H.: "Electro-thermal RF modeling and performance analysis of graphene nanoribbon interconnect," *J. Comput. Electron.* vol. 17, pp. 1695–1708, 2018.

[24] Gantmakher, V. F.: "The experimental study of electron -phonon scattering in metals," *Rep. Prog. Phys.* vol. 37, pp. 317–361, 1974.

[25] Naeemi, A., and Meindl, J. D.: "Physical modeling of temperature coefficient of resistance for single- and multi-wall carbon nanotube interconnects," *IEEE Electron Device Lett.* vol. 28, no. 2, pp. 135–138, 2007.

[26] Li, H., Srivastava, N., Mao, J.-F., Yin, W.-Y., and Kaustav, B.: "Carbon nanotube vias: Does ballistic electronphonon transport imply improved performance and reliability," *IEEE Trans. Electron Devices* vol. 58, no. 8, pp. 2689–2701, 2011.

[27] Kasap, S. O.: *Principles of Electronic Materials and Devices*, third edition, Mc-Graw Hill, New York, 2006, p. 126.

[28] Bhattacharya, S., Das, D., Rahaman, H.: "Analysis of temperature dependent power supply voltage drop in graphene nanoribbon and Cu based power interconnects," *AIMS Mater. Sci.* vol. 3, no. 4, pp. 1493–1506, 2016. doi:10.3934/matersci.2016.4.1493.

[29] Das, S., Bhattacharya, S., Das, D., and Rahaman, H.: "Thermal stability analysis of graphene nano-ribbon interconnect and applicability for terahertz frequency," *Natl. Acad. Sci. Lett.* vol. 43, no. 5, pp. 253–257, 2019.

[30] Bhattacharya, S., Das, D., and Rahaman, H.: "Modeling and performance analysis of graphene nanoribbon interconnects," *Natl. Acad. Sci. Lett.* vol. 40, no. 5, pp. 325–329.

[31] Hwang, E. H., and Sarma, S. D.: "Acoustic phonon scattering limited carrier mobility in 2D extrinsic graphene," *Phys. Rev. B* vol. 77, p. 115449, 2008,

[32] You, Y. G., Ahn, J. H., Park, B. H., Kwon, Y., Campbell, E. E. B., and Jhang, S. H.: "Role of remote interfacial phonons in the resistivity of graphene," *Applied Physics Letter* vol. 115, no. 4, 22 July 2019.

[33] Fratini, S., and Guinea, F.: "Substrate-limited electron dynamics in graphene," *Phys. Rev. B* vol. 77, p. 195415, 2008. doi:10.1103/PhysRevB.77.195415.

[34] Chen, J.-H., Jang, C., Xiao, S., Masa, and I., Fuhrer, M. S.: "Intrinsic and extrinsic performance limits of graphene devices on SiO$_2$. Nat. Nanotechnol. 2008. doi:10.1038/nnano.2008.58.

[35] Das, S., Bhattacharya, S., Das, D, and Rahaman, H.: "Modeling and analysis of electro-thermal impact of crosstalk induced gate oxide reliability in pristine and intercalation doped MLGNR interconnects," *IEEE Trans. Device Mater. Reliab.* vol. 19, no. 3, pp. 543–550, 2019.

[36] Das, S., Bhattacharya, S., Das, D., and Rahaman, H.: "A survey on pristine and intercalation doped graphene nanoribbon interconnect for future VLSI circuits," *AIMS Mater. Sci.* vol. 8, no. 2, pp. 247–260, 2021.

[37] Nishad, A. K., and Sharma, R.: "Analytical time-domain models for performance optimization of multilayer GNR interconnects," *IEEE J. Sel. Top. Quant. Electron.* vol. 20, no. 1, pp. 17–24, 2014.

[38] Bhattacharya, S., Das, S., Mukhopadhyay, A., Das, D., and Rahaman, H.: "Analysis of temperature dependent delay optimization model for GNR interconnect using wire sizing method," *J. Comput. Electron.* vol. 17, no. 4, pp. 1536–1548, 2018.

[39] Sarkar, D., Xu, C., Li, H., and Banerjee, K.: "High-frequency behavior of graphene-based interconnects-Part I: Impedance modeling," *IEEE Trans. Electron. Devices* vol. 58, no. 3, pp. 843–852, 2011.

[40] Stellari, F., and Lacatia, A. L.: "New formulas of interconnect capacitances based on results of conformal mapping method," *IEEE Trans. Electron Devices* vol. 47, no. 1, pp. 222–231, 2000.

[41] Zhao, W.-S., and Yin, W. Y.: "Comparative study on multilayer graphene nanoribbon (MLGNR) interconnects," *IEEE Electromagn. Compat. Mag.* vol. 56, no. 3, pp. 638–645, June 2014.

7 Hybrid Cu-Carbon as Interconnect Materials and Their Interconnect Models

Bhawana Kumari and Manodipan Sahoo

7.1 INTRODUCTION

Copper-carbon (Cu-carbon) hybrid is a new structure that combines copper, graphene, and CNT materials in a hybrid interconnect structure. Cu-GNR hybrid interconnects are well known to improve the reliability by adding strength to copper interconnects as a barrier layer and also improve overall conductivity by providing a parallel low-resistance path. Cu-CNT composite interconnects are also recommended to replace copper interconnects due to its higher conductivity. The proposed Cu-carbon hybrid interconnect shown in Figure 7.1 utilizes both these advantages of enhanced reliability and conductivity and emerges as a superior candidate compared to Cu-GNR and Cu-CNT composite interconnects. The structural difference between the existing interconnect structures (Cu-GNR and Cu-CNT composites) and proposed structure of Cu-carbon hybrid interconnect is shown in Figure 7.2.

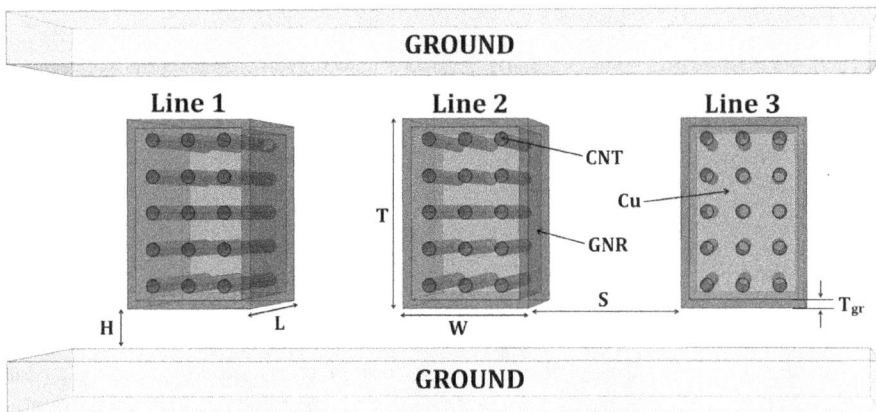

FIGURE 7.1   Three-line coupled structure of novel Cu-carbon hybrid interconnect.

DOI: 10.1201/9781003331650-7

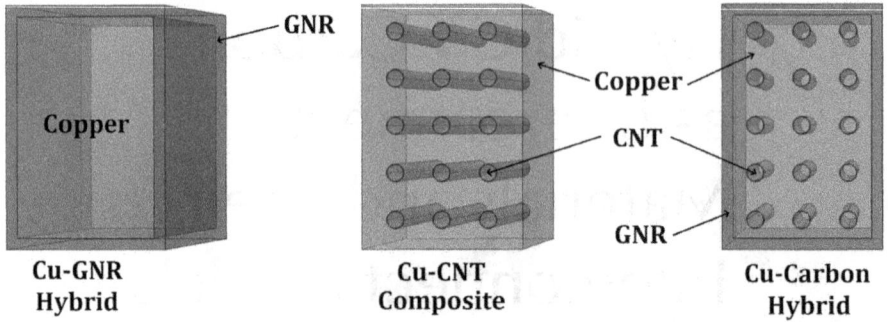

FIGURE 7.2 Structural difference between existing interconnect structures and proposed Cu-carbon hybrid interconnect structure.

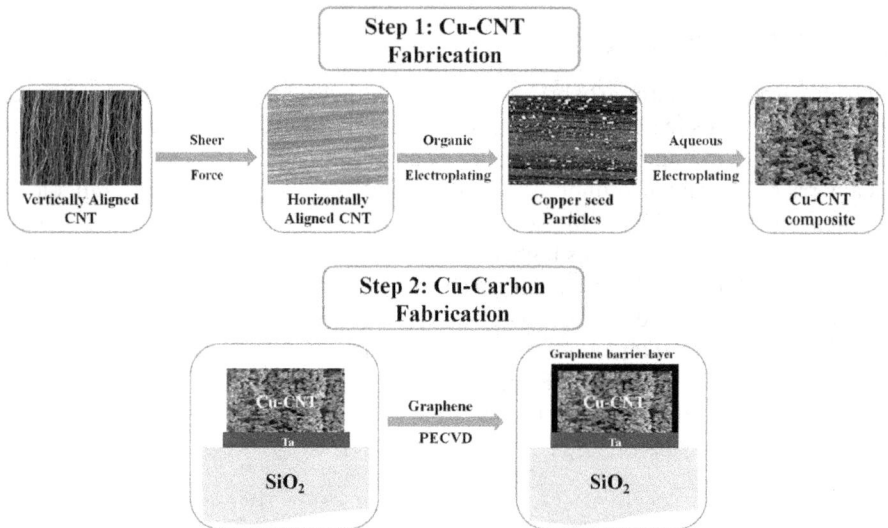

FIGURE 7.3 Schematic representation of proposed steps for Cu-carbon hybrid interconnect fabrication. Step 1: Cu seeds nucleate on the CNT surface during the organic electrodeposition and subsequently grow during the aqueous electrodeposition to yield the Cu-CNT composite [1]. Step 2: schematic of a Cu-CNT nano-wire conformally coated with graphene [2].

7.1.1 PROPOSED FABRICATION STEPS

Both Cu-GNR hybrid and Cu-CNT composites have already been fabricated and tested as reported in [1, 2, 3]. Based on that, the fabrication steps for Cu-carbon hybrid interconnect structure are proposed. A two-step fabrication process is suggested as shown in Figure 7.3. First, Cu-CNT composite is fabricated and then that Cu-CNT composite is coated with graphene, finally resulting in Cu-carbon hybrid interconnect. To fabricate a Cu-CNT composite that combines the strengths of both copper and CNT materials, copper is electrodeposited into the pores of pre-made, macroscopic CNT structures. Long, vertically aligned single-walled CNT forests can be synthesized on a substrate by the water-assisted method. The sparse forests are densified into closely packed and aligned CNT structure by the liquid densification technique with copper ions. Then, this CNT structure can be transformed into Cu-CNT composite by two-stage nucleation-growth electrodeposition process. This process is described in detail in [1]. The second step involves coating the Cu-CNT composite (formed in step 1) with graphene barrier layer. Chemical vapor deposition (CVD) is a widely used technique for graphene growth on a metal catalyst like copper. As the outer material of Cu-CNT composite material is also copper, this process will also be valid for Cu-CNT composite interconnect. A rapid low-temperature process to deposit graphene on copper nano-wire is presented in [2]. Plasma-enhanced chemical vapor deposition (PECVD) can be used to conformally coat graphene on the surfaces of Cu-CNT composite interconnect at much reduced temperatures. In brief, the carbon feedstock (methane) is dissociated into radicals using inductively coupled RF plasma at a fixed distance away from the sample holder. The carrier gas (argon) transports the reactive carbon and hydrogen radicals downstream to the Cu-CNT composite, where full graphene coverage can be achieved within 15 min at a deposition temperature of 650°C.

7.2 ELECTRICAL MODELING AND PERFORMANCE STUDY

7.2.1 First-Principle Study

The electronic properties of Cu-GNR hybrid, Cu-CNT composite, and the proposed Cu-carbon hybrid interconnects are investigated using the Density Functional Theory (DFT) approach [4]. DFT is implemented in the Atomistix Tool Kit (ATK) from Quantum Wise [5]. The atomistic structures are optimized until the maximum force of each atom becomes less than 10^{-2} eV/Å. The tolerance of the self-consistent field loop calculation is set to 10^{-4}. Figure 7.4 illustrates examples of optimized atomic structures of Cu-CNT, Cu-GNR, and Cu-carbon hybrids. The lattice orientation of copper is selected to reduce the lattice mismatch with CNT and GNR along the transport direction. We assume that CNT and GNR are not strained by copper in this study. The lattice parameters of bulk copper and armchair CNT(4,4) in DFT-GGA calculations are 3.670 and 2.474 Å, respectively. In addition, we placed copper atoms to prevent the CNT from being distorted severely. A single layer of graphene is placed on top of the atomic layers of copper. Since copper {111} plane belongs to a hexagonal lattice with nearly the same lattice constant like graphene (2% mismatch) [6], we have placed graphene atoms on {111} planes of Cu. For a VLSI interconnect, zig-zag edge (toward transport direction) of graphene nanoribbon of interest is because of its confirmed metallic electronic transport property.

Figure 7.5 shows the projected Density of States (DOS) plots for Cu-GNR hybrid, Cu-CNT composite, and Cu-carbon hybrid materials. All the materials show metallic behavior as confirmed by the energy states present at Fermi level >0 eV. Current in a 2D nano-interconnect is expressed as, $\propto \int_{\mu 1}^{\mu 2} DOS(E)\, dE$, where $(\mu_2 - \mu_1)$ is the applied potential difference [7]. Potential range of $(-0.4$ to $0.4)$ eV is considered for

Cu-GNR

Cu-CNT

Cu-Carbon

FIGURE 7.4   Atomistic structures of Cu-GNR hybrid, Cu-CNT composites, and Cu-carbon hybrid. The silver and brown spheres are carbon and copper atoms, respectively.

plotting DOS as shown in the figure. Cu-carbon hybrid exhibits larger density of energy states (lines under the graph are denser in Cu-carbon hybrid) near Fermi level when compared to other two materials as shown in the figure. Estimated current for Cu-carbon hybrid is ~21% and ~28% higher than Cu-GNR and Cu-CNT composite interconnects, respectively. This proves that Cu-carbon hybrids are more conducting than Cu-GNR and Cu-CNT composites. This deduction is supported by the circuit simulations shown in later sections (Figures 7.6–7.13).

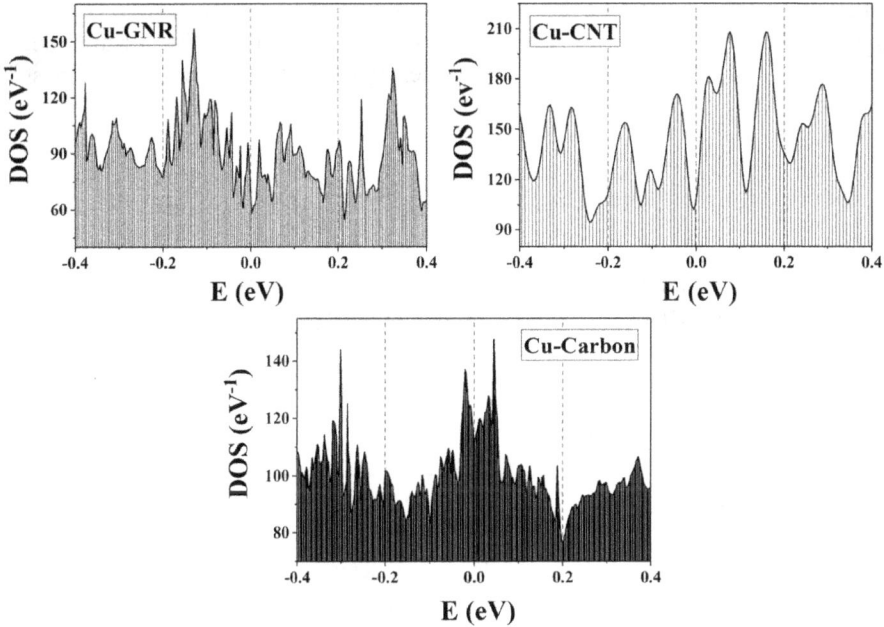

FIGURE 7.5 Projected DOS plots for (a) Cu-GNR hybrid, (b) Cu-CNT composites, and (c) Cu-carbon hybrid. Fermi level is shown by the dotted line at 0 eV.

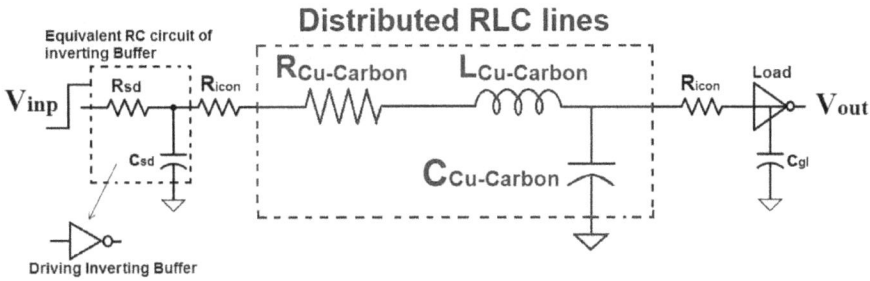

FIGURE 7.6 Equivalent electrical single metal line model for Cu-carbon hybrid interconnect structure.

7.2.2 DEVELOPMENT OF ELECTRICAL CIRCUIT MODEL OF CU-CARBON INTERCONNECTS

Here, R_{icon} denotes the imperfect contact resistance, assumed to be 10 kΩ [8]. The overall equivalent resistance of Cu-GNR hybrid interconnect considering the dielectric surface roughness is given by

$$R_{Cu-Carbon} = \left(\frac{1}{R_{Cu-CNT}} + \frac{1}{R_{GNR^{T,B}}} + \frac{1}{R_{GNR^{L,R}}} \right)^{-1} \quad (7.1)$$

where $R_{GNR}^{T,B}$ and $R_{GNR}^{L,R}$ are the resistance of the graphene barrier layer surrounding the central Cu line at the top, bottom, left, and right, respectively. The per-unit-length resistance of the multilayer graphene considering the dielectric surface roughness is given by

$$R_{GNR} = \frac{h}{2e^2 N_L N_{ch}} \left[\sum_i \left(1 + \frac{L}{\lambda_{eff^r}}\right)^{-1} \right]^{-1} \qquad (7.2)$$

N_L and N_{ch} are the number of layers and the number of channels per layer in a multilayer GNR, respectively. Length of the interconnect is represented by L. Effective mean free path of graphene nanoribbon (GNR) by accounting the scattering due to dielectric substrate surface roughness (λ_{eff}^r) is given by

$$\lambda_{eff^r} = \left(\frac{1}{\lambda_{op}} + \frac{1}{\lambda_{ac}} + \frac{1}{\lambda_{ci}} + \frac{1}{\lambda_{Spp}} + \frac{1}{\lambda_{rs}} + \frac{1}{\lambda_{edgei}} + \frac{1}{\lambda_{si}} \right)^{-1} \qquad (7.3)$$

The components in this equation are defined in detail in [9]. The per-unit-length distributed resistance of Cu-CNT composite is expressed as [10]

$$R_{Cu-CNT} = \frac{\rho_{eff}}{\omega t} \qquad (7.4)$$

where ρ_{eff} is the resistivity of Cu-carbon hybrid interconnect and is given by

$$\rho_{eff} = (1 - F_{CNT})\rho_{Cu} + F_{CNT} * \rho_{CNT} \qquad (7.5)$$

where ρ_{Cu} is the resistivity of copper, which is described as

$$\rho_{Cu} = \rho_{Cu\_bulk}(S_{FS} + S_{FS}) \qquad (7.6)$$

where ρ_{Cu_bulk} is the resistivity of bulk copper. The functions S_{FS} and S_{MS} are the Fuchs-Sondheimer and Mayadas-Shatzkes models for the surface and grain boundary scattering mechanisms in copper [10, 11]. The resistivity of SWCNT, ρ_{CNT}, is defined as [10]

$$\rho_{CNT} = \frac{\pi(D_{CNT} + 0.31 \text{nm})^2 Z_{CNT}}{4L} \qquad (7.7)$$

where Z_{CNT} is the self-impedance of an SWCNT which can be obtained by [10]

$$Z_{CNT} = \frac{L}{Nch} \left(\frac{R_{quant}}{\lambda_{eff}} \right) \qquad (7.8)$$

where λ_{eff} is the effective electron mean free path for the SWCNT and is determined by Matthiessen's equation [9].

The equivalent per-unit-length capacitance for the Cu-carbon hybrid wire to the ground plane is given by

$$C_{Cu-Carbon}^{g} = \left[\left(\sum_{j=T,B} C_{GNR}^{j} \right)^{-1} + \frac{1}{C_{CNT}^{g}} + \frac{1}{C_{e}^{g}} \right]^{-1} \tag{7.9}$$

where $(C_{GNR}^{T}, C_{GNR}^{B})$ refer to the per-unit-length quantum capacitance contributions from top and bottom graphene barrier layers surrounding Cu-CNT composite conductor, which is calculated using the recursive methodology given in [12]. C_{CNT}^{g} denotes the per-unit-length quantum capacitance contributed by SWCNT bundle calculated using the recursive methodology explained in [14]. The per-unit-length electrostatic ground capacitance is given by [12]

$$C_{e}^{g} = \varepsilon_{o}\varepsilon_{r} \left[\frac{\omega_{cu}}{h} + 2.22 \left(\frac{s}{s+0.70h} \right)^{3.19} + 1.17 \left(\frac{s}{s+1.51h} \right)^{0.76} \left(\frac{t_{cu}}{t+0.70h} \right)^{0.12} \right] \tag{7.10}$$

The equivalent per-unit-length capacitance between two Cu-carbon hybrid wires is given by

$$C_{Cu-Carbon}^{c} = \left[\left(\sum_{j=T,B} C_{GNR}^{j} \right)^{-1} + \frac{1}{C_{CNT}^{c}} + \frac{1}{C_{e}^{c}} \right]^{-1} \tag{7.11}$$

where $(C_{GNR}^{L}, C_{GNR}^{R})$ refer to the per-unit-length coupling capacitance contributions from left and right graphene barrier layers surrounding Cu-CNT composite conductor, calculated using the recursive methodology given in [12]. C_{CNT}^{c} denotes the per-unit-length coupling capacitance contributed by SWCNT bundle calculated using the recursive methodology explained in [14]. The per-unit-length electrostatic coupling capacitance is given by [12, 13]

$$C_{e}^{g} = \varepsilon_{o}\varepsilon_{r} \left[\begin{array}{c} 1.14 \frac{t_{cu}}{s} \left(\frac{h}{h+2.06s} \right)^{0.09} + 0.74 \left(\frac{\omega_{cu}}{\omega_{cu}+1.59s} \right)^{1.14} \\ +1.16 \left(\frac{\omega_{cu}}{\omega_{cu}+1.87s} \right)^{1.16} \left(\frac{h}{h+0.98s} \right)^{1.18} \end{array} \right] \tag{7.12}$$

The equivalent per-unit-length inductance of Cu-carbon hybrid interconnect is given by

$$L_{Cu-carbon} = \left(\sum_{j=T.B,L,r} \frac{1}{L_{j}} \right)_{GNR}^{-1} + L_{CNT} + L_{Cu} \tag{7.13}$$

where L_{j} is the kinetic inductance of graphene layers. L_{CNT} and L_{Cu} are per-unit-length self-inductances of CNT and copper, respectively.

TABLE 7.1

Validation of Equivalent Electrical Model with EM Solver ANSYS Q3D

Parameter	Data Obtained Using Proposed Equivalent Electrical Model	ANSYS Data	% Error
Resistance	10.14 Ω/μm	10.9 Ω/μm	6.9
Electrostatic capacitance	43.25 pF/m	47.3 pF/m	8.56
Coupling capacitance	95.252 pF/m	93.5 pF/m	1.84
Inductance	1.932 pH/m	1.95 pH/m	1

7.2.3 VALIDATION WITH EM SOLVER

The per-unit-length RLC values obtained from the equivalent electrical model described in the previous section has been verified with EM solver ANSYS Q3D. Table 7.1 shows the % error between the RLC calculated using analytical model and the EM solver.

7.2.3.1 Electrical Performance Analysis

The simulations are carried out in Cadence Virtuoso, version IC 6.1.6-64B.5004 under standard desktop environment. The distributed elements (i.e., per-unit-length R, L, and C) in Figure 7.7 are modeled with the help of 200 consecutive lumped elements [14]. The coupled three-conductor model as described in Figure 7.7 is modeled using Verilog-A including the crosstalk effects. IRDS 2018 roadmap [15] is considered for extracting the parameters used in the calculations at 7 nm technology node. Worst-case crosstalk delays for Cu, Cu-GNR hybrid, Cu-CNT composite, and Cu-carbon hybrid interconnects are observed and compared in Figure 7.8 at 7 nm node. Cu-carbon hybrid interconnect experiences the least crosstalk delay as it has the lowest per-unit-length resistance. It is also observed that Cu-GNR hybrid has better performance than Cu-CNT composite interconnect. Delay is lesser by ~31%, ~23%, and ~8% for 100-μm-long; ~88%, ~28%, and ~41% for 1-mm-long; and ~86%, ~31%, and ~68% for 5-mm-long Cu-carbon hybrid interconnects when compared to Cu, Cu-GNR, and Cu-CNT interconnects, respectively.

Figure 7.9 shows the time-domain response of Cu-GNR, Cu-CNT, and Cu-carbon hybrid interconnects. As observed, Cu-carbon hybrid interconnect has the lowest switching delay and its step response is steeper as compared to others. The capacitance of Cu-carbon hybrid, which is the dominating factor, is least among all (capacitance in Cu-carbon is a series combination of capacitances in Cu-CNT and GNR) and hence the steepest step response. The step response of Cu-GNR hybrid interconnect is steeper (because of lower time constant) as compared to Cu-CNT composite interconnect.

With victim net (i.e., middle line in Figure 7.7) at logic high level, if aggressor nets (i.e., upper and lower lines in Figure 7.7) switch from logic low to high, then overshoot occurs in the victim net. But, with the victim net kept at logic low

Distributed RLC lines

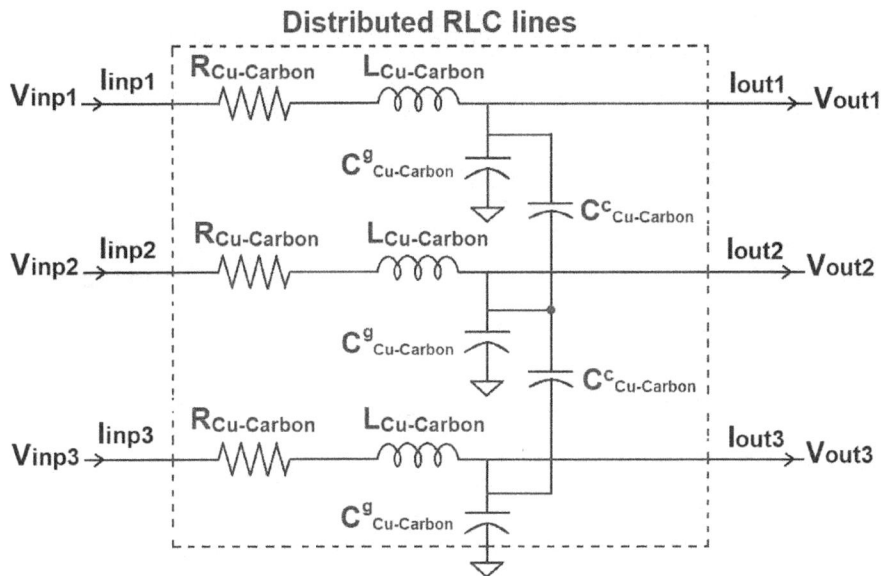

FIGURE 7.7 Electrical equivalent of three-line Cu-carbon composite interconnect system.

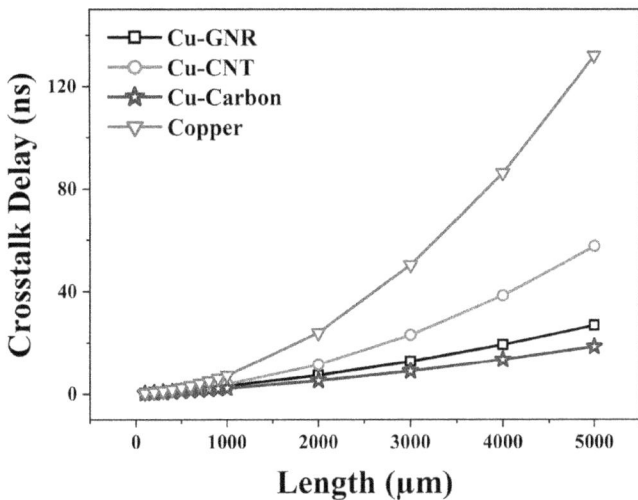

FIGURE 7.8 Worst-case crosstalk-induced delay in Cu-GNR hybrid, Cu-CNT composites, and Cu-carbon hybrid interconnects.

level, if both the aggressor nets switch from logic high to low, then undershoot occurs. Noise-peak represents the measure of the overshoot and undershoot voltage occurring in the signal transmitted through interconnect [14]. ABCD parameter matrix model is utilized for the analysis of crosstalk-induced noise effects [12].

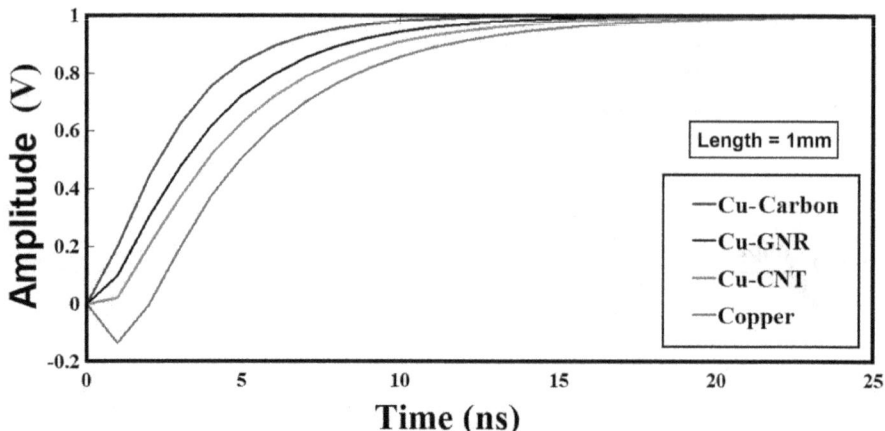

FIGURE 7.9 Time-domain response of Cu-GNR hybrid, Cu-CNT composites, and Cu-carbon hybrid interconnects. Length of interconnects is considered to be 1 mm.

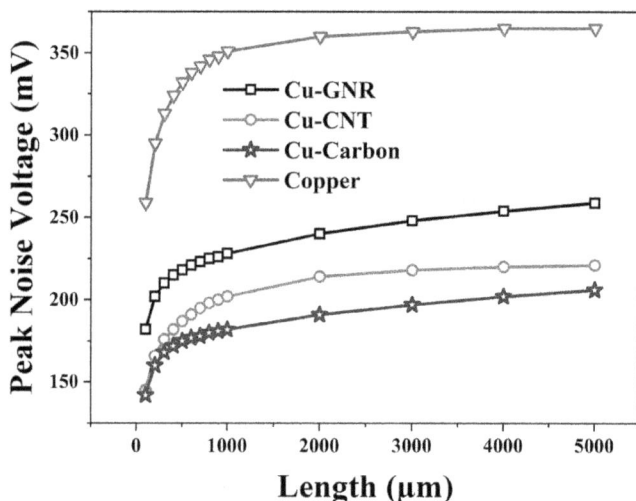

FIGURE 7.10 Peak-noise voltage in Cu-GNR hybrid, Cu-CNT composites, and Cu-carbon hybrid interconnects.

The noise-peak voltages induced in victim nets due to crosstalk for Cu-GNR hybrid, Cu-CNT composite, and Cu-carbon hybrid interconnects are shown in Figure 7.10. The peak crosstalk noise voltage depends upon the coupling ratio (C.R.), which is a function of coupling capacitance. Cu-carbon hybrid interconnect has the least coupling capacitance among all, hence lowest noise-peak voltage is observed. Cu-CNT composite has lower noise-peak as compared to Cu-GNR hybrid because of higher time constant.

Noise-delay-product (NDP) is defined as the product of worst-case crosstalk delay and noise-peak voltage. It expresses the measure of signal integrity of the

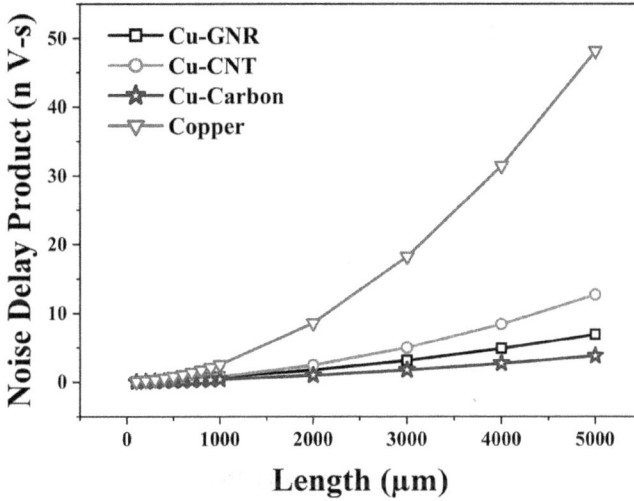

FIGURE 7.11   Noise-delay-product in Cu-GNR hybrid, Cu-CNT composites, and Cu-carbon hybrid interconnects.

victim metal line. Figure 7.11 shows comparison of noise-delay-product for Cu-GNR hybrid, Cu-CNT composite, and Cu-carbon hybrid interconnects. The trend is similar to delay as it is the dominating factor in NDP. NDP is lesser by ~62%, ~40%, and ~10% for 100-μm-long; ~84%, ~42%, and ~47% for 1-mm-long; and ~92%, ~45%, and ~70% for 5-mm-long Cu-carbon hybrid interconnects when compared to Cu, Cu-GNR, and Cu-CNT interconnects, respectively. Cu-carbon hybrid interconnect outperforms other alternative interconnect structures in terms of signal integrity, making it the most suitable candidate.

Repeaters are inserted in long distributed wires to linearize delay with respect to interconnect length. The calculation of optimal number of repeaters (N_{opt}) and optimal size of repeaters (S_{opt}) (i.e., ratio between driver size and load size) is explained in detail in [16]. The total power consumed in the interconnect is mainly because of the power consumed by driver and load buffers, which is given by [16]

$$P_{total} = \left(N_{opt} + 2\right)\left(P_{switch} + P_{short} + P_{leak}\right) \tag{7.14}$$

where P_{switch}, P_{short}, and P_{leak} are switching, short circuit, and leakage power of a repeater, respectively. The definitions of various parameters are specified in detail in [16].

Figure 7.12 shows power consumption in Cu-GNR hybrid, Cu-CNT composite, and Cu-carbon hybrid interconnects. Although number of repeaters is high in copper interconnects the repeater size is the least among all, decreasing its switching power, leading to least power consumption among all alternatives. Cu-carbon hybrid consumes lesser power when compared to Cu-GNR hybrid and Cu-CNT composite interconnects. Although the number of repeaters is low in Cu-GNR hybrid interconnect, the required repeater size is largest, implying highest power consumption. Power-Delay-Product (PDP) is defined as the product of worst-case crosstalk delay and power consumption.

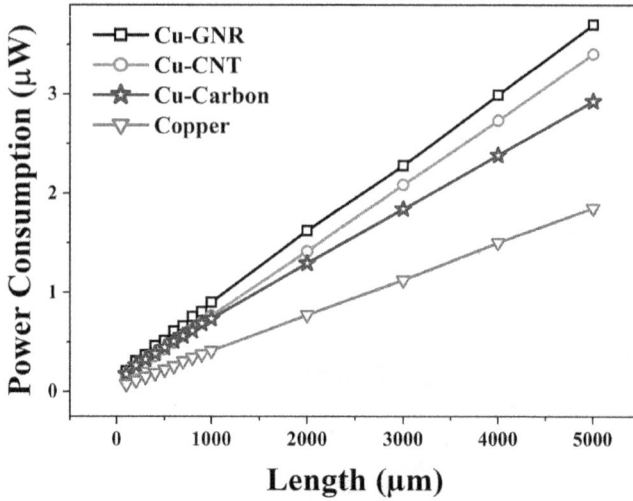

FIGURE 7.12   Power consumption in Cu-GNR hybrid, Cu-CNT composites, and Cu-carbon hybrid interconnects.

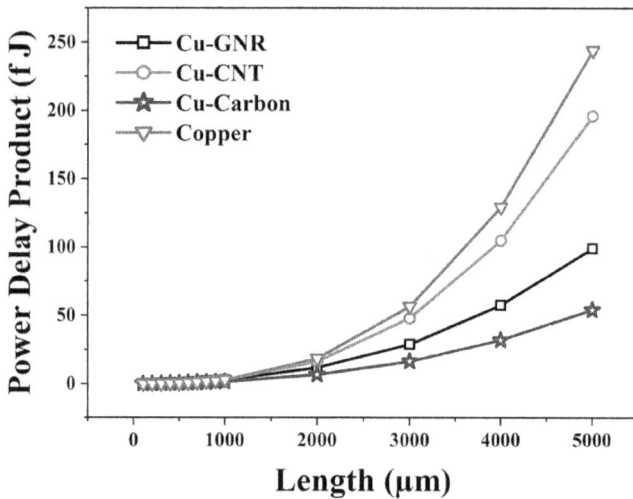

FIGURE 7.13   Power-delay-product in Cu-GNR hybrid, Cu-CNT composites, and Cu-carbon hybrid interconnects.

It depicts a tradeoff between performance and power for the victim metal line. Cu-carbon hybrid experiences least PDP, making it a suitable candidate for both high-performance and low-power applications. Even though Cu-GNR hybrid consumes higher power, its PDP is lower than Cu-CNT composite interconnect as delay plays a major role in PDP calculation. PDP is lesser by ~39%, ~34%, and ~13% for 100-μm-long; ~43%, ~41%, and ~44% for 1-mm-long; and ~78%, ~46%, and ~72% for 5-mm-long Cu-carbon hybrid interconnects when compared to Cu, Cu-GNR hybrid, and Cu-CNT composite interconnects, respectively. From the point of view of both VLSI circuit performance and power dissipation, our proposed Cu-carbon hybrid interconnect proves to be a superior candidate for future VLSI applications.

7.3 HIGH-FREQUENCY MODELING AND PERFORMANCE STUDY

7.3.1 DEVELOPMENT OF AC ANALYTICAL MODELING OF CU-CARBON HYBRID INTERCONNECTS

The equivalent distributed per-unit-length (p.u.l) resistance of Cu-carbon hybrid is defined as

$$R_{Cu-Carbon} = Re\left(\frac{1}{\sigma_{eff}}\right)\frac{1}{WT} \tag{7.15}$$

where σ_{eff} represents the conductivity of Cu-carbon hybrid, which can be defined as

$$\sigma_{eff} = (1 - F_{CNT})\sigma_{Cu} + F_{CNT} * \sigma_{CNT} + \sigma_{GNR} \tag{7.16}$$

where σ_{Cu} is the conductivity of copper as described in [17]. The conductivity of CNT ($\sigma_{cnt} = \dfrac{4L}{\pi(D_{cnt} + 0.31\text{nm})^2 Z_{cnt}}$) and GNR ($\sigma_{gnr} = \dfrac{L}{WT\sum_{j=t,b,l,r} Z_{gnr}}$) is expressed as [7], where Z_{cnt} and Z_{gnr} are intrinsic self-impedance of an isolated SWCNT and an isolated MLGNR, which are expressed as [7]

$$Z_{cnt} = \frac{L}{N_{ch}^c}\left(\frac{R_{quant}}{\lambda_{eff}^c} + jwL_K\right) \tag{7.17}$$

$$Z_{gnr} = \frac{L}{N_{ch}^g N_L}\left(\frac{R_{quant}}{\lambda_{eff}^g} + jwL_K\right) \tag{7.18}$$

The p.u.l equivalent capacitance of Cu-carbon hybrid interconnect can be expressed as

$$C_{Cu-Carbon} = \left[\left(\sum_{j=t,b,l,r} C_{gnr}^j\right)^{-1} + \frac{1}{C_{cnt}} + \frac{1}{C_e}\right]^{-1} \tag{7.19}$$

The p.u.l equivalent inductance of Cu-carbon hybrid interconnect consists of internal and external inductances, which are described as

$$L_{Cu-Carbon} = L_{in} + L_{ex} = Im\left(\frac{1}{\sigma_{eff}}\right) + \frac{\mu_o \varepsilon_o \varepsilon_r}{C_e} \tag{7.20}$$

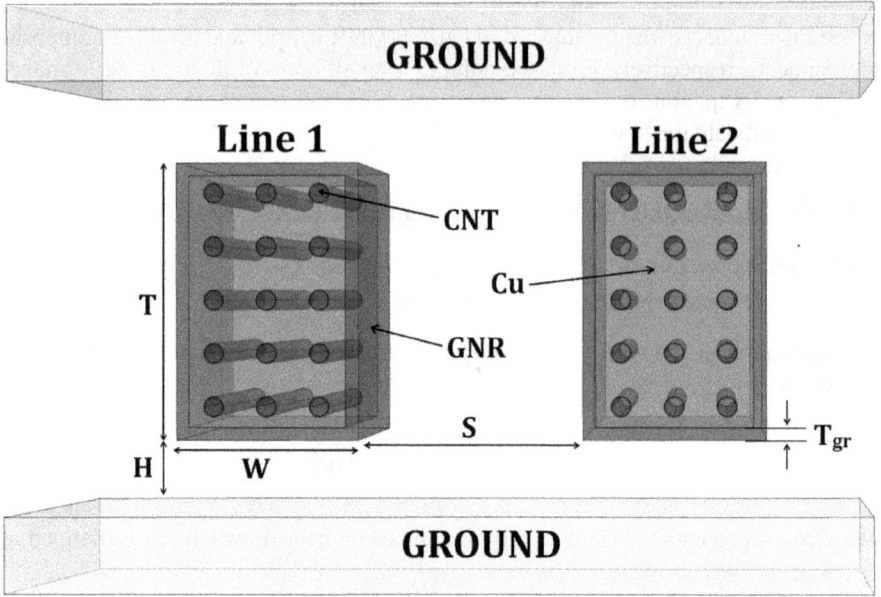

FIGURE 7.14   Two-line coupled Cu-carbon hybrid interconnect structure.

Figure 7.15 demonstrates the electrical equivalent circuit of a 2-line coupled Cu-carbon hybrid interconnect system. Here, $C^c_{Cu-Carbon}$ and $L^m_{Cu-Carbon}$ represent the p.u.l coupling capacitance and mutual inductance developed between the Cu-carbon hybrid metal lines, respectively, which are taken from [19]. The effective complex conductivity of Cu-carbon hybrid interconnect is calculated by substituting $Z_{cnt} = R_{cnt} + j\omega\, L_{cnt}$, $Z_{gnr} = R_{gnr} + j\omega\, L_{gnr}$ and (7.16) into (7.33), one can obtain

$$\sigma_{eff} = (1 - F_{cnt})\sigma_{Cu} + \left(\frac{k_1 F_{cnt} R_{cnt}}{R^2_{cnt} + \omega^2 L^2_{cnt}} + \frac{k_2 R_{gnr}}{R^2_{gnr} + \omega^2 L^2_{gnr}} \right)$$
$$- j \left(\frac{k_1 F_{cnt} R_{cnt}}{R^2_{cnt} + \omega^2 L^2_{cnt}} + \frac{k_2 R_{gnr}}{R^2_{gnr} + \omega^2 L^2_{gnr}} \right) \tag{7.21}$$

and the ratio between $Re(\sigma_{eff})$ and $Im(\sigma_{eff})$ is described by

$$\left| \frac{Re(\sigma_{eff})}{Im(\sigma_{eff})} \right| = \frac{(1 - F_{cnt})\sigma_{Cu}(R^2_{cnt} + \omega^2 L^2_{cnt})(R^2_{gnr} + \omega^2 L^2_{gnr}) + k_1 F_{cnt} R_{cnt}(R^2_{gnr} + \omega^2 L^2_{gnr}) + k_2 R_{gnr}(R^2_{gnr} + \omega^2 L^2_{gnr})}{\omega(k_1 F_{cnt} L_{cnt}(R^2_{gnr} + \omega^2 L^2_{gnr}) + k_2 L_{gnr}(R^2_{gnr} + \omega^2 L^2_{gnr}))}$$
$$\tag{7.22}$$

At first, $|\sigma_{eff}|$ decreases with increasing frequency, then saturates at higher frequencies as can be observed from equation (7.21). Equation (7.22) expresses the modulus of $|Re(\sigma_{eff})/Im(\sigma_{eff})|$. This equation states the fact that the reactive part Cu-CNT composite interconnects is dominant at higher frequencies while that is not the case with Cu-carbon hybrid interconnects. This dominance reduces with increase in F_{cnt}.

7.3.1.1 ABCD Parameter-Based Single Line Model

The equivalent electrical model shown in Figure 7.6 can be simplified as a transmission line whose transmission characteristics can be expressed by utilizing an ABCD matrix [11, 13, 20 21]. The total ABCD transmission parameter matrix of the configuration shown in Figure 7.6 is defined by equation (1.23). The final expressions for A, B, C, and D parameters for an isolated interconnect line are defined by equations (7.24–7.27).

$$
\begin{bmatrix} A & B \\ C & D \end{bmatrix} = \begin{bmatrix} 1 & R_{sd} \\ 0 & 1 \end{bmatrix} \begin{bmatrix} 1 & 0 \\ sC_{sd} & 1 \end{bmatrix} \begin{bmatrix} 1 & R_{icon} \\ 0 & 1 \end{bmatrix}
$$

$$
\begin{bmatrix} \cosh(\varphi h) & Z_o \sinh(\varphi h) \\ \dfrac{1}{Z_o}\sinh(\varphi h) & \cosh(\varphi h) \end{bmatrix} \begin{bmatrix} 1 & R_{icon} \\ 0 & 1 \end{bmatrix} \begin{bmatrix} 1 & 0 \\ sC_{gl} & 1 \end{bmatrix}
$$

$$(7.23)$$

$$
A = (1 + sC_{gl}R_{icon}) \left[\begin{array}{l} (1 + sC_{sd}R_{sd})\cosh(\varphi L) + \dfrac{1}{Z_o}(R_{icon} + R_{sd}) \\[2mm] + sC_{sd}R_{icon}R_{sd})\sinh(\varphi L) \end{array} \right]
$$

$$
+ s \left[\begin{array}{l} C_{gl}(1 + sC_{sd}R_{sd})Z_o \sinh(\varphi L) + (R_{icon} \\[2mm] + R_{sd} + sC_{sd}R_{icon}R_{sd} \cosh(\varphi L)) \end{array} \right]
$$

$$(7.24)$$

$$
B = (1 + sC_{sd}R_{sd})R_{icon}\cosh(\varphi L) + \dfrac{R_{icon}}{Z_o}(R_{icon} + R_{sd} + sC_{sd}R_{icon}R_{sd})\sinh(\varphi L)
$$

$$
+ (1 + sC_{sd}R_{sd})Z_o \sinh(\varphi L) + (R_{icon} + R_{sd} + sC_{sd}R_{icon}R_{sd} \cosh(\varphi L))
$$

$$(7.25)$$

$$
C = (1 + sC_{gl}R_{icon}) \left[sC_{sd}\cosh(\varphi L) + \dfrac{1}{Z_o}(1 + sC_{sd}R_{icon})\sinh(\varphi L) \right]
$$

$$
+ sC_{gl}(sC_{sd}Z_o \sinh(\varphi L) + (1 + sC_{sd}R_{icon}\cosh(\varphi L)))
$$

$$(7.26)$$

$$
D = sC_{sd}R_{icon}\cosh(\varphi L) + \dfrac{R_{icon}}{Z_o}(1 + sC_{sd}R_{icon})\sinh(\varphi L)
$$

$$
+ sC_{sd}Z_o \sinh(\varphi L) + (1 + sC_{sd}R_{icon}\cosh(\varphi L))
$$

$$(7.27)$$

where Z_0 and ϕ are the characteristic impedance and the propagation constant of single line interconnect. These parameters can be expressed as

$$
Z_0 = \sqrt{\dfrac{(R_{Cu-carbon} + s(L_{Cu-Carbon}))}{s(\alpha C_{Cu-Carbon})}}
$$

$$(7.28)$$

$$
\varphi = \sqrt{s(\alpha C_{Cu-Carbon})(R_{Cu-carbon} + s(L_{Cu-Carbon}))}
$$

$$(7.29)$$

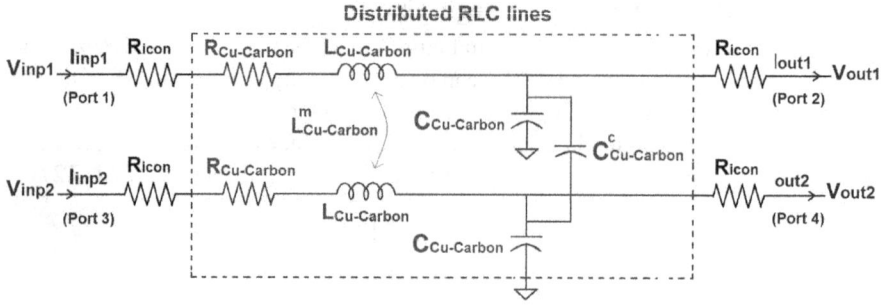

FIGURE 7.15 Equivalent electrical 2-line coupled system model for Cu-carbon hybrid interconnect system.

7.3.1.2 ABCD Parameter-Based Coupled Line Model

Similarly, for a 2-line coupled interconnect system configuration as shown in Figure 7.15, the total ABCD transmission parameter matrix can be expressed by

$$
\begin{bmatrix} A_{11} & A_{12} & B_{11} & B_{12} \\ A_{21} & A_{22} & B_{21} & B_{22} \\ C_{11} & C_{12} & D_{11} & D_{12} \\ C_{21} & C_{22} & D_{21} & D_{22} \end{bmatrix} = \begin{bmatrix} 1 & 0 & R_{sd} & 0 \\ 0 & 1 & 0 & R_{sd} \\ 0 & 0 & 1 & 0 \\ 0 & 0 & 0 & 1 \end{bmatrix} \begin{bmatrix} 1 & 0 & 0 & 0 \\ 0 & 1 & 0 & 0 \\ sC_{sd} & 0 & 1 & 0 \\ 0 & sC_{sd} & 0 & 1 \end{bmatrix} \begin{bmatrix} 1 & 0 & R_{icon} & 0 \\ 0 & 1 & 0 & R_{icon} \\ 0 & 0 & 1 & 0 \\ 0 & 0 & 0 & 1 \end{bmatrix}
$$
$$
[CC^n] \begin{bmatrix} 1 & 0 & R_{icon} & 0 \\ 0 & 1 & 0 & R_{icon} \\ 0 & 0 & 1 & 0 \\ 0 & 0 & 0 & 1 \end{bmatrix} \begin{bmatrix} 1 & 0 & 0 & 0 \\ 0 & 1 & 0 & 0 \\ sC_{gl} & 0 & 1 & 0 \\ 0 & sC_{gl} & 0 & 1 \end{bmatrix}
$$

(7.29)

where CCn is the ABCD matrix representing a 2-line coupled system constructed using n number of cascaded infinitesimal sections and is given by

$$
[CC^n] = \begin{bmatrix} \frac{1}{2}(\cosh(\rho_1 L) + \cosh(\rho_2 L)) & \frac{1}{2}(\cosh(\rho_1 L) - \cosh(\rho_2 L)) & \frac{1}{2}(Z_1 \sinh(\rho_1 L) + Z_2 \sinh(\rho_2 L)) & \frac{1}{2}(Z_1 \sinh(\rho_1 L) - Z_2 \sinh(\rho_2 L)) \\ \frac{1}{2}(\cosh(\rho_1 L) - \cosh(\rho_2 L)) & \frac{1}{2}(\cosh(\rho_1 L) + \cosh(\rho_2 L)) & \frac{1}{2}(Z_1 \sinh(\rho_1 L) - Z_2 \sinh(\rho_2 L)) & \frac{1}{2}(Z_1 \sinh(\rho_1 L) + Z_2 \sinh(\rho_2 L)) \\ \frac{1}{2}(Z_1 \sinh(\rho_1 L) + Z_2 \sinh(\rho_2 L)) & \frac{1}{2}(Z_1 \sinh(\rho_1 L) - Z_2 \sinh(\rho_2 L)) & \frac{1}{2}(\cosh(\rho_1 L) + \cosh(\rho_2 L)) & \frac{1}{2}(\cosh(\rho_1 L) - \cosh(\rho_2 L)) \\ \frac{1}{2}(Z_1 \sinh(\rho_1 L) - Z_2 \sinh(\rho_2 L)) & \frac{1}{2}(Z_1 \sinh(\rho_1 L) + Z_2 \sinh(\rho_2 L)) & \frac{1}{2}(\cosh(\rho_1 L) - \cosh(\rho_2 L)) & \frac{1}{2}(\cosh(\rho_1 L) + \cosh(\rho_2 L)) \end{bmatrix}
$$

(7.30)

7.3.2　AC Performance Analysis

The modeling and simulations are executed in MATLAB software, version R2021b under standard desktop environment. Validation of the proposed model is performed in Advanced Design System (ADS) software. The diameter of a SWCNT in a bundle is considered to be 1 nm. The CNT filling ratio (F_{cnt}) in the hybrid structure is taken as 0.6. Considering global level interconnects, width and length of the line are taken as 15 and 1 mm, respectively. Thickness of MLGNR barrier layer is considered as

2.5 nm, which corresponds to six layers of graphene. All the parameters used here in calculations are extracted from IRDS 2017 roadmap [18] at 10 nm technology node.

Figure 7.16 shows the AC magnitude vs frequency plot of Cu-carbon hybrid, copper, and Cu-CNT composite interconnects. Cu-carbon hybrid interconnect has the highest bandwidth (BW) due to its least resistance among other alternatives.

Figure 7.17a and b shows the frequency-dependent resistances and inductances of Cu-carbon hybrid, copper, and Cu-CNT composite metal lines, respectively. The resistances of the metals increases while inductances decrease with increasing frequency. Cu-carbon hybrid interconnects have the lowest impedance among all. By incorporating CNT bundles in copper metal line, the impedance can be suppressed,

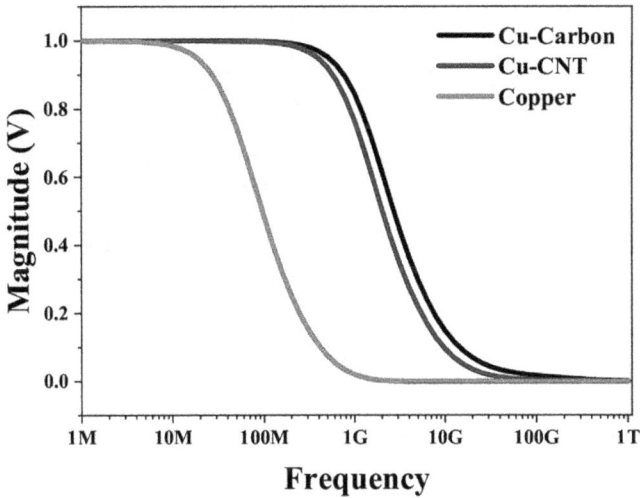

FIGURE 7.16   AC magnitude vs frequency plot of Cu-carbon hybrid, Cu-CNT composite, and copper interconnects vs frequency.

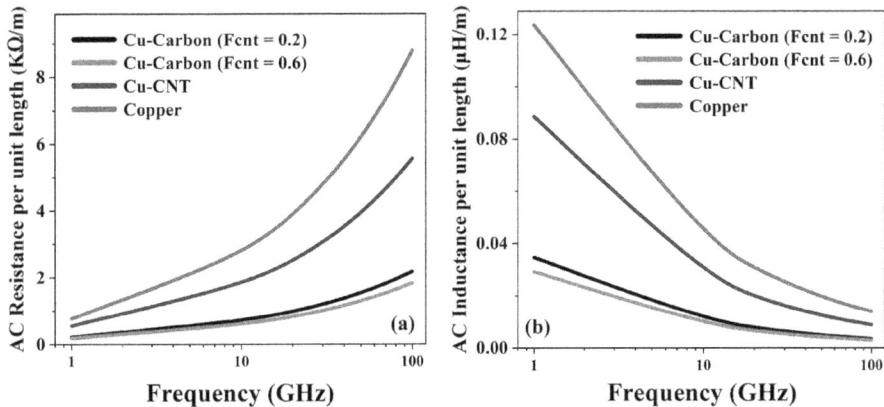

FIGURE 7.17   (a) Per-unit-length AC resistance and (b) per-unit-length AC inductance of Cu-carbon hybrid, copper, and Cu-CNT composite interconnects.

resulting in improved performance in the composite as compared to copper interconnect. As expected, the resistance of Cu-carbon hybrid increases with the decrease
of F_{cnt}. The magnitudes (in dB) of return loss coefficient S_{11} for single and 2-line
coupled interconnects made of Cu-carbon hybrid, Cu-CNT composite, and copper
are plotted in Figures 7.18 and 7.19, respectively.

S_{11} value increases with increase in frequency. Cu-carbon hybrid encounters lowest return loss when compared to other alternatives. Cu-carbon hybrid interconnect

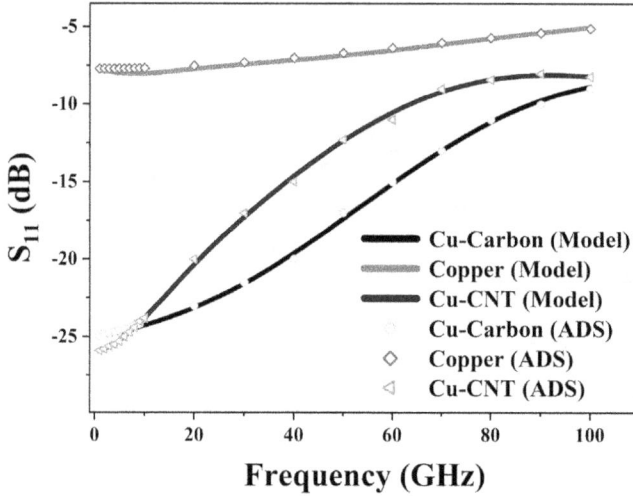

FIGURE 7.18   Return loss coefficient (S_{11}) of single Cu-carbon hybrid, copper, and Cu-CNT
composite interconnect lines. Validation with ADS software is reflected through dotted points.

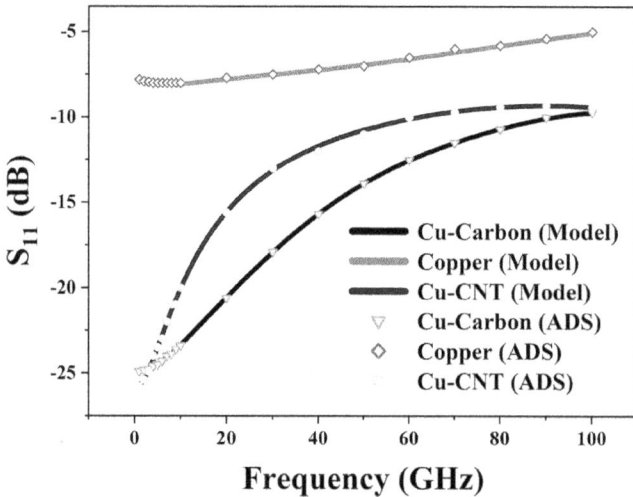

FIGURE 7.19   Return loss coefficient S_{11} of 2-line coupled Cu-carbon hybrid, copper, and
Cu-CNT composite interconnect lines. Validation with ADS software is reflected through
dotted points.

has ~68% lower S_{11} value than copper at lower frequencies but even at 100 Hz frequency, the improvement is ~43% for single line and ~48% for 2-line coupled interconnects. Data points estimated from ADS simulator match well with the plots obtained from our model as shown in the figures. The magnitudes of forward transmission coefficient S_{21} of single line and 2-line coupled interconnects made of Cu-carbon hybrid, copper, and Cu-CNT composite are demonstrated in Figures 7.20 and 7.21, respectively. S_{21} value decreases with increase in frequency.

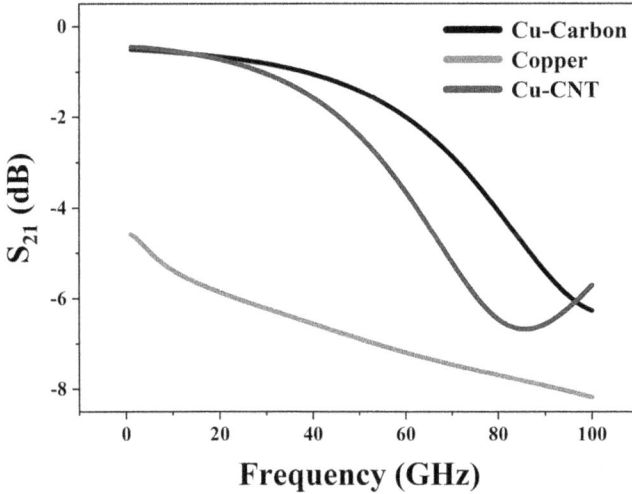

FIGURE 7.20 Forward transmission coefficient (S_{21}) of single Cu-carbon hybrid, copper, and Cu-CNT composite interconnect lines.

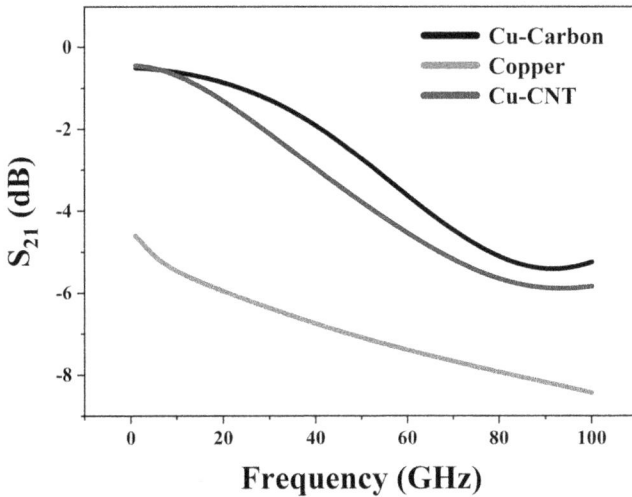

FIGURE 7.21 Forward transmission coefficient (S_{21}) of 2-line coupled Cu-carbon hybrid, copper, and Cu-CNT composite interconnect lines.

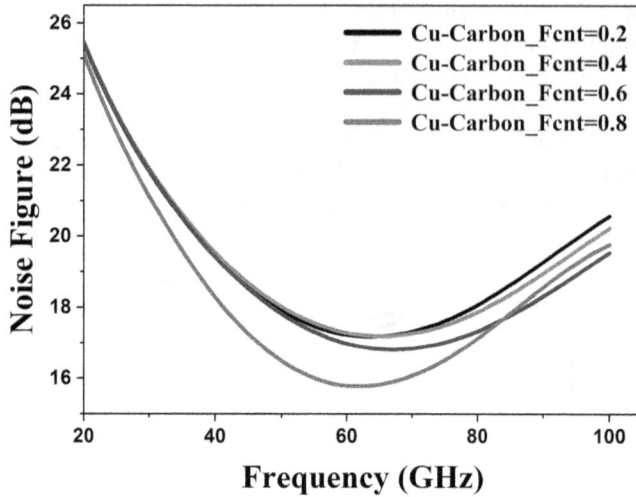

FIGURE 7.22 Noise figure corresponding to input at port 1 and output from port 4 of 2-line coupled Cu-carbon hybrid interconnect for different F_{cnt}.

Single line Cu-carbon hybrid interconnect structure has higher bandwidth for gain below −3 dB. Cu-carbon hybrid interconnect has highest transmission coefficient. At 100 GHz, Cu-carbon hybrid interconnect has ~30% and ~38% higher S21 values than copper for single line and 2-line coupled interconnects, respectively.

The effect of F_{cnt} on the noise parameter of Cu-carbon hybrid interconnects is depicted in Figure 7.22. This study tells us that a CNT fraction of 0.8 and 0.6 is a good choice for high-frequency noise-constrained applications. $F_{cnt}=0.6$ is selected for further evaluations because it performs better at certain range of frequencies and also it has been experimentally demonstrated by Sundaram et al. [3].

The subsequent noise analysis is done for 2-line coupled interconnect systems to study the effects of crosstalk and interference in high-frequency applications by applying input at port 1 of line 1 and observing output at port 4 of line 2. Figure 7.23 gives an understanding of the noise figure in the 2-line coupled interconnect system. At lower frequencies, all interconnects have comparable noise profiles as observed in Figure 7.23. Noise in copper remains almost constant with increasing frequency. Cu-CNT and Cu-carbon hybrid interconnect experiences a steep dip in noise figure followed with a gradual rise. The percentage improvement (in dB) in the noise figure of Cu-carbon hybrid interconnect as compared to copper is ~48% at 100 GHz. The noise factor corresponding to input at port 1 and output from port 4 for 2-line coupled interconnect lines is demonstrated in Figure 7.24. As seen from the figure, the degradation in signal to noise ratio in Cu-carbon hybrid interconnect is far better as compared to copper with an improvement of ~98% at 100 GHz. The improvement is ~44% when compared to Cu-CNT composite interconnect.

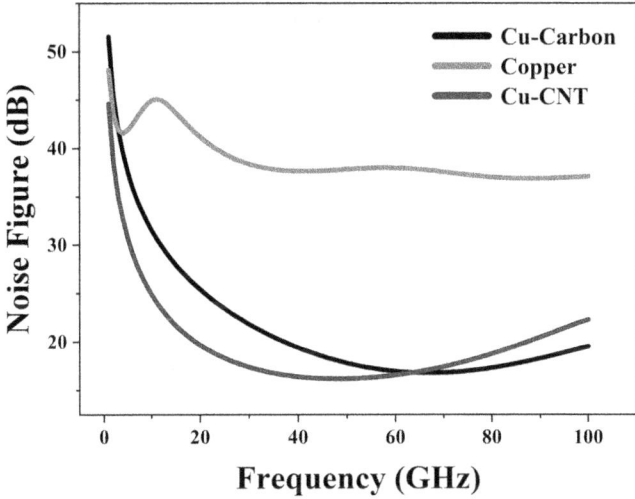

FIGURE 7.23 Noise figure corresponding to input at port 1 and output from port 4 for 2-line coupled Cu-carbon hybrid, copper, and Cu-CNT composite interconnect lines.

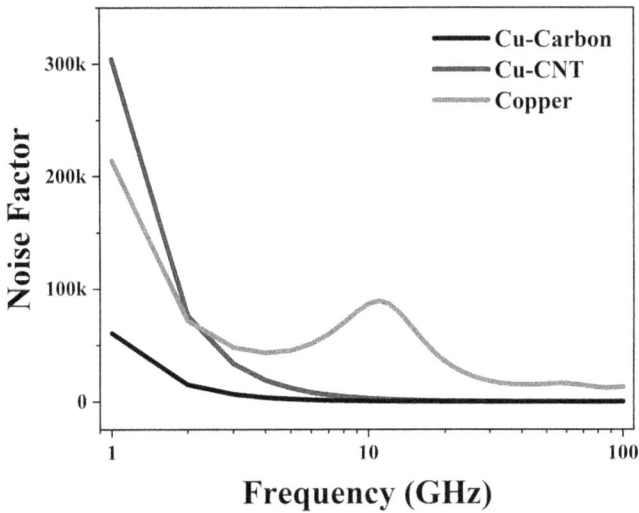

FIGURE 7.24 Noise factor corresponding to input at port 1 and output from port 4 for 2-line coupled Cu-carbon hybrid, copper, and Cu-CNT composite interconnect lines.

7.4 ELECTRO-THERMAL STUDY

7.4.1 Development of Electro-Thermal Model of Cu-Carbon Hybrid Interconnects

Finite element method (FEM) is very popular for characterizing the hybrid effects of electrical and thermal effects in various 3D structures. The methodology used to perform this study can be described with the help of following electro-thermal model. The current continuity and heat diffusion equations in the steady state are given by

$$\nabla . J_{in} = \nabla . \left(\sigma_{eff} \nabla \rho \right) = 0 \tag{7.31}$$

$$\nabla \cdot (\kappa \nabla T) = -q \tag{7.32}$$

where φ is the potential. The heat source can be given as $q = JE$, E represents the electric field. σ_{eff} denotes the effective electrical conductivity of Cu-carbon hybrid, which can be defined as

$$\sigma_{eff} = (1 - F_{cnt})\sigma_{Cu} + F_{cnt}\sigma_{cnt} + \sigma_{gnr} \tag{7.33}$$

where $\sigma_{Cu}(T)$ represents the temperature-dependent effective conductivity of copper metal line,

$$\sigma_{Cu}(T) = \frac{\sigma_0}{1 + \xi(T - T_{amd})(\rho_{fs} + \rho_{gbs})} \tag{7.34}$$

Here, σ_0 is the bulk resistivity of copper metal line at ambient temperature, $T_{amb} = 293$ K, $\xi = 0.003862$ K^{-1} is the temperature coefficient of resistivity, ρ_{fs} and ρ_{gbs} are the functions representing the surface and grain boundary scattering mechanisms in copper taken from the Fuchs-Sondheimer and Mayadas-Shatzkes models [10],

$$\rho_{fs} = 0.45 \Lambda_{Cu} (1 - P_{Cu}) \frac{W_{cu} + T_{h,cu}}{W_{cu} T_{h,cu}} \tag{7.35}$$

$$\rho_{gbs} = \left[1 - 1.5x + 3x^2 - 3x^3 \ln\left(1 + \frac{1}{x}\right) \right]^{-1} \tag{7.36}$$

$$x = \frac{\Lambda_{Cu}\beta_{ref}}{S_g(1 - \beta_{ref})} \tag{7.37}$$

Λ_{Cu} is the mean free path of copper, P_{Cu} is the secularity parameter, W_{cu} and $T_{h,cu}$ are the width and thickness of copper wire, respectively. β_{ref} is the reflection coefficient at the grain boundary and S_g is the grain size of copper. The conductivity of SWCNT and GNR is expressed as

$$\sigma_{cnt} = \frac{4L(\eta_{ch}(T)\Lambda_{eff}^e(T))}{\pi(D_{cnt} + 0.31\text{nm})^2 R_{quant}}, \sigma_{gnr} = \frac{L(\eta_{ch}(T)\eta_L \Lambda_{eff}^e(T))}{WT_h \displaystyle\sum_{j=t,b,l,r} R_{quant}} \tag{7.40}$$

where R_{quant} denotes the quantum resistance of CNT and graphene materials [7]. η_{ch} is the number of conducting channels in one shell of SWCNT and one layer of MLGNR, which is given by

$$\eta_{ch}(T) = \sum_{all\_subbands} \frac{1}{\exp(|E_v| / K_B T) + 1} \tag{7.41}$$

$\Lambda^e_{mfp}(T)$ denotes the effective electron mean free path for the ith shell of SWCNT and jth layer of MLGNR, which can be evaluated by Matthiessen's equation [10]:

$$\Lambda^e_{mfp}(T) = \left(\frac{1}{\Lambda_{AC}} + \frac{1}{\Lambda_{OP,ems}} + \frac{1}{\Lambda_{op,abs}} \right) \qquad (7.42)$$

7.4.2 ELECTRO-THERMAL PERFORMANCE AND ANALYSIS

Two types of structures are utilized in this work (i.e., via-interconnect-via and BEOL structures). The BEOL is organized in different metal layers, local (M_x), intermediate, semi-global, and global wires. The total number of layers can be as many as 15, while the typical number of M_x layers ranges between 3 and 6. Each of these layers contains (unidirectional) metal lines organized in regular tracks and dielectric materials. They are interconnected vertically by means of via structures that are filled with metal. A signal generated by the MOSFET propagates through the BEOL to reach the receiver MOSFET on a chip. The simulations are executed in COMSOL multiphysics software under standard desktop environment. Different combinations of materials (such as copper, Cu-CNT, Cu-GNR, and Cu-carbon) for interconnects and vias are studied here. Considering intermediate level interconnects for via-interconnect-via structure as shown in Figure 7.25, width, aspect ratio, and length of the line are taken as 15 nm, 2.0, and 50 μm, respectively. The diameter of a SWCNT in a bundle is considered to be 1 nm. The CNT filling ratio (F_{cnt}) in the hybrid structure is taken as 0.6. Thickness of MLGNR barrier layer is considered as 2.5 nm, which corresponds to six layers of graphene. All the parameters used here in calculations are extracted from IRDS 2017 roadmap [22] at 10 nm technology node. The magnitude of the current that M_x lines take from the FEOL

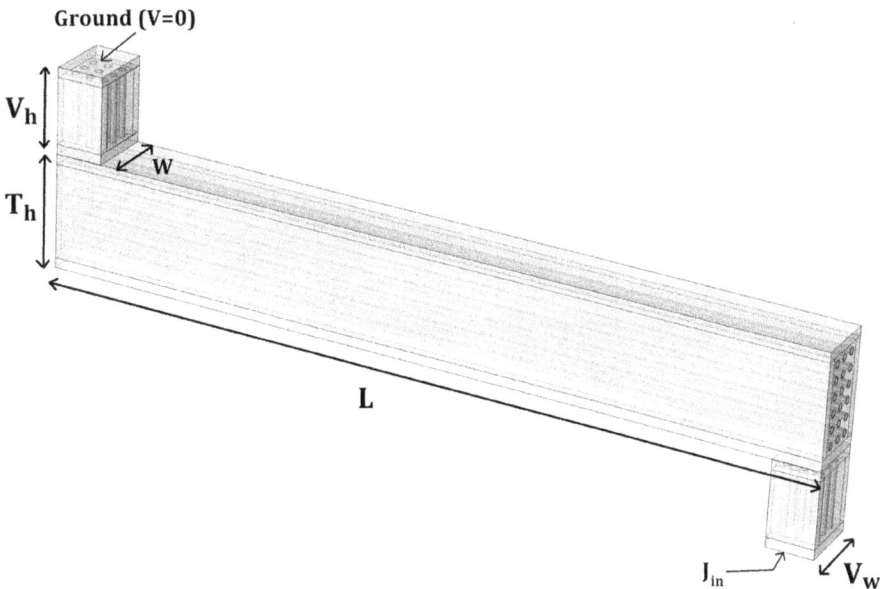

FIGURE 7.25 Structural representation of copper, CNT, and novel Cu-carbon hybrid interconnect structure.

can be approximated as a saturated drain to source current for a basic elemental block of all construction (CMOS Inverter logic) [27]. This can be given as a current terminal at the metal line surfaces in FEM simulator. The via-interconnect-via structure is loaded by a dc current of density J_{in} (MA/cm²) at the bottom of one via and the other end is grounded to observe the electro-thermal behavior of different materials. For studying the Joule heating of the BEOL, heat diffusion equation can be solved assuming a constant heat flux source at the FEOL (bottom plate). The power density of FEOL is given as $P_T = 1000$ W/m. COMSOL utilizes Joule heating multiphysics, which takes into account the properties of the electric currents and the conductive heat transfer properties of the materials.

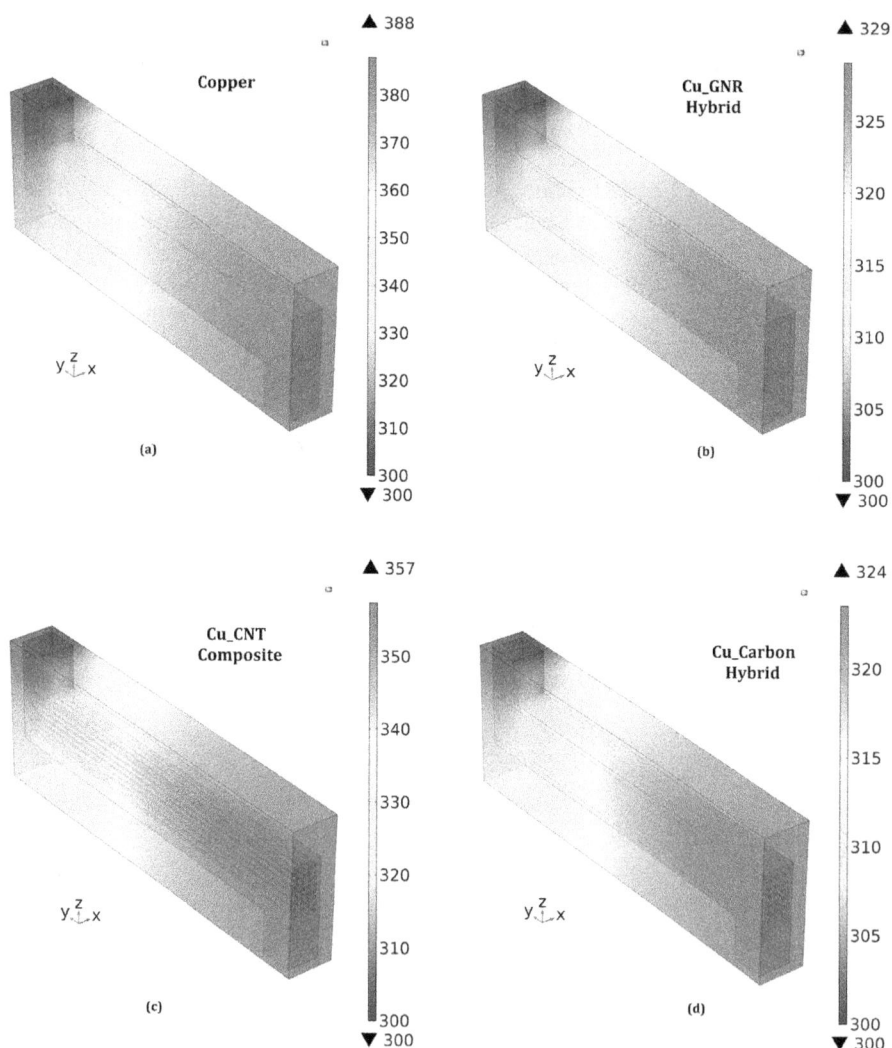

FIGURE 7.26 Temperature profiles of different via-interconnect-via structures. (a) All copper, (b) copper-Cu-GNR-copper, (c) copper-Cu-CNT-copper, and (d) copper-Cu-carbon-copper.

Figure 7.26 shows the temperature profiles of different via-interconnect-via structures. Copper heats up to highest temperature due to self-heating as compared to other alternative structures. Cu-carbon hybrid experiences the lowest temperature rise among others.

The maximum temperature attained by Cu-carbon is less than that of copper by 16%. In Figure 7.27, the temperature profile of all Cu-carbon structure is shown. By replacing copper with Cu-carbon via, the rise in temperature is reduced by 4°C. The temperature profiles for two BEOL structures are shown in Figure 7.28. An additional current density is applied at the boundary of one of the metal lines in level 2 to analyze the heat distribution. Uniform distribution of heat can be observed in Cu-carbon hybrid in addition to low temperature rise. But copper-based BEOL exhibits hotspot formation at metal level 2, indicating low thermal efficiency as compared to Cu-carbon hybrid-based BEOL.

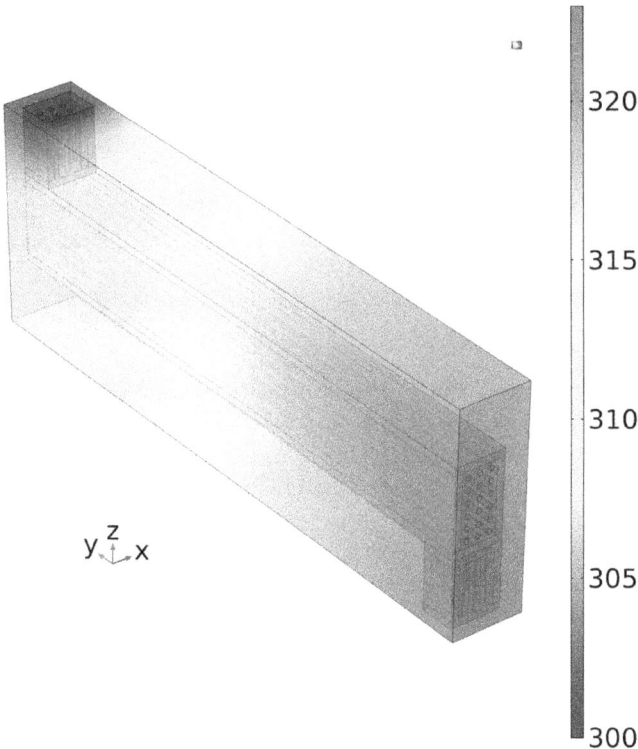

FIGURE 7.27 Temperature profiles for all Cu-carbon via-interconnect-via structure.

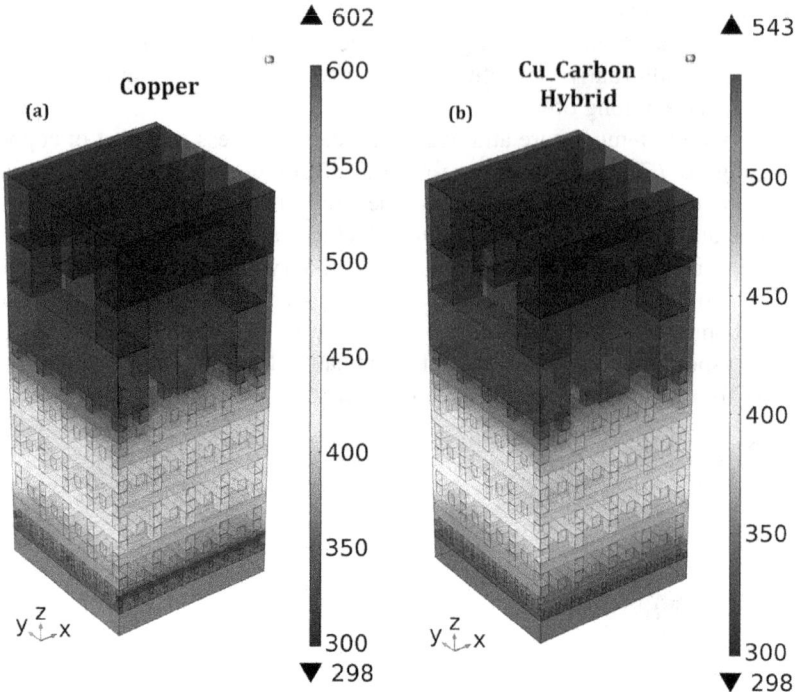

FIGURE 7.28 Temperature profiles of BEOL structures. (a) Copper-based BEOL and (b) Cu-carbon-based BEOL.

7.5   CONCLUSION

This study proposes for the first time a new structure combining copper, CNT, and GNR materials into a hybrid structure. This new structure synergistically combines the strengths of all the three materials (i.e., Cu, CNT, and GNR). The steps required to fabricate this structure are also proposed by utilizing the fabrication methods of Cu-CNT composite and Cu-GNR hybrid materials. The atomistic simulations suggest that Cu-carbon hybrid structure is more conductive than its parent structures, i.e., Cu-CNT composite and Cu-GNR hybrid. This is also supported by the circuit simulation results which show that Cu-carbon hybrid interconnect experiences lowest delay among all other alternatives. Time-domain analysis also suggests that Cu-carbon hybrid interconnect has the steepest and sharpest step response. Crosstalk-induced noise is least in Cu-carbon hybrid interconnects. Cu-carbon hybrid interconnect has proven to be better than other alternatives in terms of signal integrity. Power consumption is also least in Cu-carbon hybrid interconnects. For the first time, a high-frequency electrical model of single and 2-line coupled Cu-carbon hybrid metal lines is proposed and various transmission parameters are analyzed at 10 nm technology node. The high-frequency performance of Cu-carbon hybrid is studied and compared with existing copper and Cu-CNT composite interconnects. The ABCD parameter model for single and 2-line coupled interconnects is developed, which are utilized to obtain the scattering parameters. This model is also validated with ADS software.

Here, Cu-carbon hybrid interconnects have the lowest impedance among all, which tends to increase with the decrease of F_{cnt}. When compared with copper, Cu-carbon hybrid interconnect (with $F_{cnt}=0.6$) possesses $\sim$80% lower impedance at 100 GHz frequency. The reactive part of Cu-CNT composite interconnects is dominant at higher frequencies while that is not the case with Cu-carbon hybrid interconnects where both real and imaginary parts contribute equally in the effective complex conductivity. Cu-carbon hybrid interconnect experiences lowest return loss and highest forward transmission gain among all others. At lower frequencies, all interconnects have comparable noise profiles. Noise figure (in dB) and noise factor of Cu-carbon hybrid interconnect when compared to copper show $\sim$48% and $\sim$98% improvement at 100 GHz, respectively. The improvement in noise factor is $\sim$44% when compared to Cu-CNT composite interconnect. Even though Cu-carbon and Cu-CNT have comparable noise figures, still Cu-carbon is recommended due to its lower AC impedances and better signal transmission. Cu-carbon hybrid is an emerging interconnect structure with very promising performance metric and phenomenal advantages over copper interconnects. An electro-thermal model of Cu-carbon hybrid interconnect is proposed here for the first time. Temperature profiles of Cu-carbon hybrid interconnect corresponding to input current source are simulated and compared to existing interconnect alternatives, copper, Cu-CNT composite, and Cu-GNR hybrid interconnects. Copper interconnects heat up to highest temperature due to self-heating as compared to other alternative structures while Cu-carbon hybrid experiences the lowest temperature rise among others. The maximum temperature attained by Cu-carbon is less than that for copper by 16%. Even in the influence of an additional current source at metal level 2, uniform distribution of heat can be observed in Cu-carbon hybrid BEOL in addition to low temperature rise. Whereas copper-based BEOL exhibits hotspot formation at metal level 2 indicating low thermal efficiency as compared to Cu-carbon hybrid-based BEOL. Cu-carbon hybrid is an emerging interconnect structure with very promising performance metric and phenomenal advantages over copper interconnects. These findings conclude that our proposed Cu-carbon hybrid interconnect is a superior candidate for future VLSI applications. Therefore, it needs to be studied and analyzed extensively in theory and experiment for various aspects of VLSI applications.

REFERENCES

[1] R. M. Sundaram, A. Sekiguchi, M. Sekiya, T. Yamada, and K. Hata, "Copper/carbon nanotube composites: Research trends and outlook", *R. Soc.* vol. 5, no. 11, p. 180814, 2018.
[2] C. Subramaniam, T. Yamada, K. Kobashi et al., "One hundred fold increase in current carrying capacity in a carbon nanotube-copper composite", *Nat. Commun.* vol. 4, p. 2202, 2013.
[3] R. Mehta, S. Chugh, and Z. Chen, "Enhanced electrical and thermal conduction in graphene-encapsulated copper nanowires", *Nano Lett.* vol. 15, no. 3, pp. 2024–2030, 2015.
[4] M. Brandbyge, J.-L. Mozos, P. Ordejn, J. Taylor, and K. Stokbro, "Density-functional method for nonequilibrium electron transport", *Phys. Rev. B* vol. 65, p. 165401, 2002.
[5] Atomistix ToolKit version 2016.4, *QuantumWise A/S*, https://www.quantumwise.com, Accessed December 4, 2017.
[6] M. M. Khoshdel, E. Targholi, and M. J. Momeni, "First-principles calculation of quantum capacitance of codoped graphenes as supercapacitor electrodes", *J. Phys. Chem. C* vol. 119, no. 47, pp. 26290–26295, 2015.

[7] S. Dutta, *Quantum Transport*, Cambridge University Press, Cambridge, 2005.

[8] J.-O. Lee, C. Park, J.-J. Kim, J. Kim, J. W. Park, and K.-H. Yoo, "Formation of low-resistance ohmic contacts between carbon nanotube and metal electrodes by a rapid thermal annealing method", *J. Phys. D Appl. Phys.* vol. 33, no. 16, pp. 1953–1956, 2020.

[9] R. Kumar, B. Kumari, S. Kumar, M. Sahoo, and R. Sharma, "Temperature and dielectric surface roughness dependent performance analysis of Cu-graphene hybrid interconnects", In: *2020 IEEE Electrical Design of Advanced Packaging and Systems (EDAPS)*, IEEE, Shenzhen, China, 2020, pp. 1–3, doi: 10.1109/EDAPS50281.2020.9312905.

[10] Z. H. Cheng, W. S. Zhao, L. Dong et al., "Investigation of copper-carbon nanotube composites as global VLSI interconnects", *IEEE Trans. Nanotechnol.* vol. 16, no. 6, pp. 891–900, 2017.

[11] M. Sahoo and H. Rahaman, "Modeling of crosstalk induced effects in copper-based nanointerconnects: An ABCD parameter matrix-based approach", *J. Circ. Syst. Comput.* vol. 24, no. 2, pp. 218–1266, 2015.

[12] M. Sahoo and H. Rahaman, "Modeling and analysis of crosstalk induced overshoot/undershoot effects in multilayer graphene nanoribbon interconnects and its impact on gate oxide reliability", *Microelectron. Reliab.* vol. 63, pp. 231–238, 2016.

[13] M. Sahoo and H. Rahaman, "Modeling of crosstalk induced effects in copper-based nanointerconnects: An ABCD parameter matrix-based approach", *J. Circ. Syst. Comput.* vol. 24, no. 2, pp. 218–1266, 2015.

[14] M. Sahoo, P. Ghosal, and H. Rahaman, "Modeling and analysis of crosstalk induced effects in multiwalled carbon nanotube bundle interconnects: An ABCD parameter-based approach", *IEEE Trans. Nanotechnol.* vol. 14, no. 2, pp. 259–274, 2015.

[15] IRDS, *International Roadmap for Devices and Systems (IRDS-2018) Reports*, https://irds.ieee.org/editions/2018.

[16] M. Sahoo, H. Rahaman, and B. Bhattacharya, "On the suitability of single-walled carbon nanotube bundle interconnects for high-speed and power-efficient applications", *J. Low Power Electron.* vol. 10, pp. 479–494, 2014.

[17] M. Ahmadloo and A. Dounavis, "Parameterized model order reduction of electromagnetic systems using multiorder Arnoldi", *IEEE Trans. Adv. Pack.* vol. 33, no. 4, pp. 1012–1020, 2010.

[18] Z. H. Cheng, W. S. Zhao, D. W. Wang et al., "Analysis of Cu-graphene interconnects", *IEEE Access* vol. 6, pp. 53499–53508, 2018.

[19] W. Zhao, X. Li, S. Gu, S. H. Kang, M. M. Nowak, and Y. Cao, "Field-based capacitance modeling for sub-65-nm on-chip interconnect", *IEEE Trans. Electron Devices* vol. 56, no. 9, pp. 1862–1872, 2009.

[20] D. Fathi and B. Forouzandeh, "Time domain analysis of carbon nanotube interconnects based on distributed RLC model", *NANO* vol. 4, no. 1, pp. 13–21, 2019.

[21] M. Sahoo, P. Ghosal, and H. Rahaman, "An ABCD parameter based modeling and analysis of crosstalk induced effects in multiwalled carbon nanotube bundle interconnects", In: *2014 27th International Conference on VLSI Design and 2014 13th International Conference on Embedded Systems*, IEEE, Mumbai, India, 2014, pp. 433–438, doi: 10.1109/VLSID.2014.81.

[22] IRDS, *International Roadmap for Devices and Systems (IRDS-2017) Reports*, https://irds.ieee.org/editions/2017.

8 Hybrid Cu-CNT Composite as Interconnect Materials and Their Equivalent Models

Shivangi Chandrakar and Manoj Kumar Majumder

8.1 INTRODUCTION

In order to integrate the modules on a piece of silicon, recent VLSI designs require millions of closely spaced interconnects. Owing to its compatibility with silicon, aluminum and tungsten have been utilized as interconnect material for many years. However, with scaling down of device dimensions, reliability worsens as a result of an increased current density that can lead to electromigration-related failures. Therefore, the interconnect material is replaced by copper that possesses a lower resistance than aluminum and tungsten counterparts [1]. The Cu-based interconnect modeling and integration have been the subject of numerous investigations. A higher integration density and design complexity related to the Cu-based interconnect is becoming one of the primary challenges for continued scaling of device dimensions. Cu was subsequently found to be inappropriate for high-speed interconnects for the below reasons:

1. The increasing surface and grain boundary scatterings at the nanoscale regime considerably reduces the electrical conductivity of Cu.
2. Electromigration-induced hillocks and voids are promoted by the increasing joule heating and resistivity of copper.

Thus, the Cu interconnect undergone through well-established electrochemical deposition (ECD) method creates good conductivity and thermal stability. The difficulties of electromigration, interconnect temperature, and extremely diffusive barrier layer are the key defects that emerged with its multiple benefits and made it a hard alternative for a high aspect ratio through silicon via (TSV) interconnect in 3D IC [2]. However, by introducing graphene nano-ribbons (GNR) and carbon nanotubes (CNT) as a filler material in via, these issues may be readily overcome. A CNT/GNR exhibits exceptional features such as strong current conduction ability, longer mean

DOI: 10.1201/9781003331650-8

free pathways (MFPs) at lower biasing condition, and an improved electro-thermal stability [3,4]. However, the fabrication and other relevant production process for CNT/GNR-based interconnects are costly and immature still. Therefore, researchers are working toward the production of Cu-CNT/GNR-based hybrid interconnect technology to support the present silicon-based industry. In the following subsections, the advantage and uniqueness of Cu-CNT-based hybrid interconnect are explored in comparison with the other emerging interconnect materials.

8.1.1 CNT/GNR-Based Interconnects

Single-walled carbon nanotubes (SWCNTs) and multi-walled carbon nanotubes (MWCNTs) (shown in Figure 8.1) have piqued the research community's interest in the recent decades, due to their remarkable tensile competences, high thermal conductivity, and unique electrical characteristics (ballistic transport) [5–7]. The sp^2 carbon-carbon bond is the stiffest and strongest in nature and it approaches maximum tensile stiffness and strength. Experimental analysis has revealed that CNT possesses a high Young's modulus of greater than 1 TPa and a strength of approximately 100 GPa that is much higher than the other metallic compositions and carbon fibers. The density of CNT is roughly 1.75 g/cm^3, which is relatively low and found to have a thermal conductivity of 6,000 W/m K, which is a pretty good value. The long ballistic nature of CNT can achieve the current density of 10^{11} Am^{-2} that is greater than copper counterparts [8–12]. Furthermore, GNR is a single sheet of graphene layer that is exceptionally thin and narrow, resulting in a 1D structure [13]. As a consequence, GNRs may be thought of as an unrolled form of CNTs, with most of their electronic characteristics being comparable to CNTs. In terms of manufacture, it is clear that the GNR's planar form makes it easier to regulate its development than that of the CNT. The planar structure makes it compatible for the traditional lithography process [14].

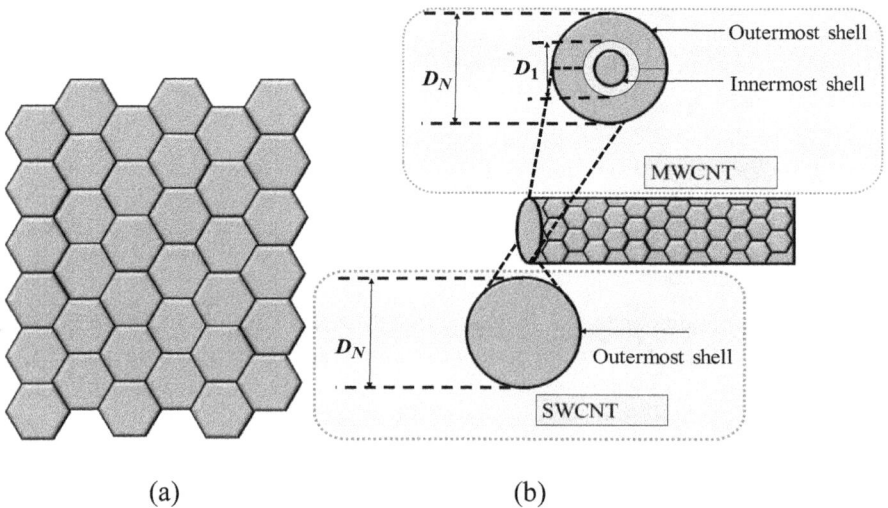

(a)   (b)

FIGURE 8.1 Sketch representing graphene sheet and CNT as (a) unrolled graphene sheet and (b) rolled carbon nanotube.

8.1.2 SILICON NANOWIRE-BASED INTERCONNECTS

As a result of their unique physical and electrical characteristics, silicon nanowires (SiNWs) (shown in Figure 8.2) are being explored for a potential interconnect material. Eventually, they may be utilized to join nanoscale components in ultra-compact electronic devices. SiNWs are often referred to be solid, cylindrical-shaped wires that have diameters in the nanometer regime. A procedure known as chemical vapor deposition is used for the production of semiconductor nanowires. Nanowires of this kind may be created using a technique known as vapor-liquid-solid (VLS) or solution-liquid-solid (SLS). The VLS development procedure is considered as the approach with the greatest degree of adaptability. The VLS growth method relies on the employment of a metal, typically an Au nanoparticle, to catalyze NW. The nanowires are produced as a byproduct of the chemical vapor deposition process, which involves the decomposition of source gases such as $SiCl_4$ [15]. Gold is the metal that works best for nanowire development, since it is made from rare earth elements and is expensive. Therefore, it is not often employed in the bulk production process of semiconductors. As a result, it is essential to discover some additional catalysts to choose. Copper is one of the potential options; however, it can only be used at high temperatures. There are a number of different physical considerations that point to the potential of a nanowire's conductivity being lower than that of the analogous bulk material. The scattering process that occurs close to the wire borders is one of the primary reasons for this. This scattering process has a significant influence on the lower wire width values, which is typically lower than the electron mean free path (*mfp*) of the bulk material. Nevertheless, the scattering nature might not be seen in CNTs because the electrons' mobility at this carbon tubes follows the rules of ballistic transport. This makes it difficult to see the scattering effect in the CNT material. As a result, nanowire seems to have a high electrical resistance and is useful for applications such as IC resistances and electronics sensor [16].

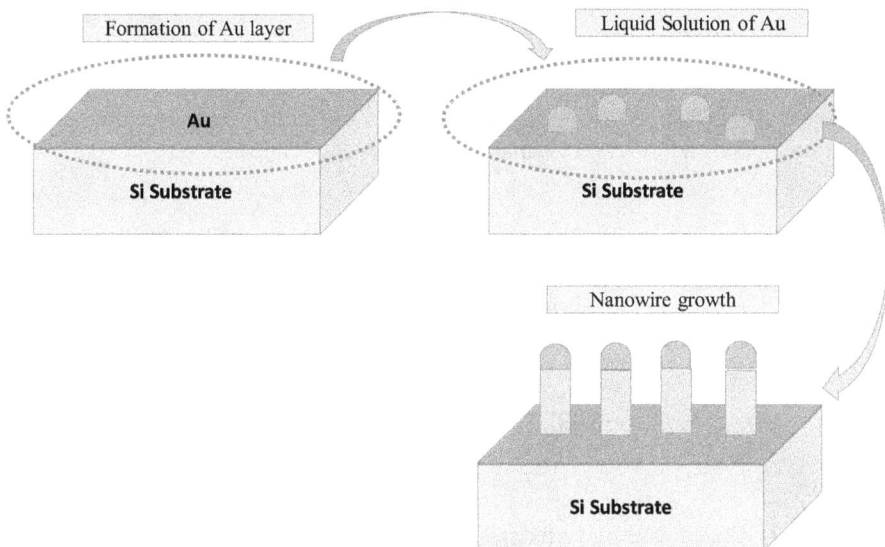

FIGURE 8.2 Formation of silicon nanowire in silicon substrate.

8.1.3   PLASMONIC-BASED INTERCONNECTS

Plasmonic combines the large bandwidth provided by photonics with the higher integration density of the nanostructures. This is accomplished by connecting the energy of a photon with the energy of a free electron gas that results in the creation of an oscillating nature termed as a plasmon. Plasmonic mechanisms are able to achieve subwavelength confinement that precisely serves as the foundation of the nanophotonics study. In recent years, plasmonic-based devices have been confronted with major hurdles as a result of losses that have been experienced in optoelectronics. Due to the significant losses, use of the aforementioned materials in a wide range of innovative applications is severely hindered. The movement of electrons is the primary factor that determines how devices that are based on plasmonics function. Metals are unable to maintain a homogeneous distribution of their electrons when exposed to higher frequencies. As a result, many of the electrons are packed into a single location that may be interpreted as having an extremely negative charge, while the other locations are seen as having a positive charge. According to the theory of electromagnetics, electric field lines are able to pass from positively charged locations to negatively charged locations, and this phenomenon happens all across the interconnect wires. If a voltage is provided all over the terminals of the interconnect, the field lines will travel at the speed of light along the interface between both the metal and the dielectric in the instance of the applied voltage. Herein, the plasmons are separated into two categories: surface plasmon polaritons (SPP) and localized surface plasmons (LSP), and their distinction is determined by the electrons' agitation found inside the metal. Surface electromagnetic (EM) wave is said to exhibit SPP behavior if it travels in a path that is parallel to the metal-dielectric interface [17,18]. The collective electron charge oscillations known as LSP may be seen in metallic nanoparticles when they are activated by light (shown in Figure 8.3). The Ag, Au, Al, graphene, and other materials are used in the fabrication of the metal components of the plasmonic technology. Metals such as silver and gold are effective for both low- and high-frequency surface plasmon propagation (LSP and SSP), while graphene is an even more effective plasmonic material for THz frequency. When compared to more traditional metal/dielectric interfaces, graphene's advantageous qualities—including its distinctive band structure, better carrier transport, and lower loss—make it the material of choice.

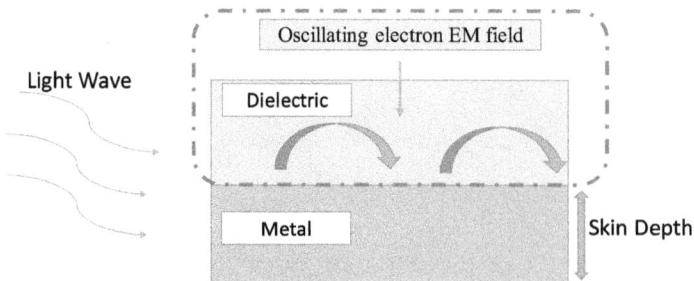

FIGURE 8.3   Optical effect of surface plasmon polaritons in the metal-dielectric junction.

8.1.4 SPINTRONIC-BASED INTERCONNECTS

Dimensional scaling has been accompanied by massive heat production during the last several decades. This might also restrict Moore's law at lower technological node. Post-CMOS mechanism, which functions with diverse state variables including electron spin, has received the greatest attention because of its benefits in terms of stability, non-volatility, and improved performance. Spintronic is a conventional term for the study of spin electronics, which is a branch of physics. The electron spin may be regulated by an outer magnetic field and the polarization of the electrons themselves in the field of spintronics. The electric current can be regulated by the employment of these polarized electrons. The manipulation of the quantum mechanical feature of electrons known as the spin of electrons is the foundation of the field of spintronics that derives its name from this attribute. In the near future, when electrons have a spin degree of freedom, this will provide future electronic devices a substantial improvement in their versatility and efficiency. Electrons, being negatively charged subatomic particles, may store and transport information in digital devices. In digital electrical equipment, the existence or nonexistence of electrons inside a substrate or equivalent material signifies the zeroes and ones of processor binary data. However, in the area of spintronics, the data is stored and conveyed utilizing alternative feature of electrons known as spin. Spin represents the inherent electron angular momentum, and all the electron functions resembling a miniature bar magnet, pointing either upward or downward to symbolize an electron's spin. The consequence of the cumulative magnetic field impact is zero owing to the mobility of an electron having random spin. It is possible to store binary data in the form of ones—all spin ups and zeroes, i.e., all spins down using external magnetic fields. The phenomena were primarily seen in a system that was constructed out of alternating layers of magnetic and nonmagnetic materials that were electrically conducting. Using a magnetic field, the electron spin shifted from all up to all down, achieving it the label "spin valve" [19,20].

8.1.5 COPPER-CARBON NANOTUBES COMPOSITE: THE WORTHIEST OPTION

In the nanoscale regime of VLSI industry, it should be emphasized that even with latest fabrication aspects, the assumption of tightly packed CNTs is incorrect. The analysis has been found that a SWCNT bundle possesses a maximum density of $1.5 \times 10^{13} cm^2$ [21]. The CNT filling ratio (f_{CNT}) is stated as the fraction of CNTs to overall area, and it is found to be just 0.2, with a conductivity substantially lower than the Cu counterpart [22]. A bundle of copper-carbon nanotube (Cu-CNT) composite has been created in order to increase electrical properties without sacrificing reliability. The Cu-CNT combination has a similar conductivity but a 100 times greater ampacity compared to the Cu material. A Cu-CNT composite also exhibits strong electromigration resistance [23,24]. In all other respects, the Cu-CNT composite can establish a reasonable trade-off between performance and reliability. As a result, it stands to reason that the effective properties of Cu-CNT composites as a chip interconnect can be the primary reason for the consideration in the latest VLSI interconnect modeling.

8.2 FABRICATION ASPECT OF HYBRID CU-CNT COMPOSITE-BASED ON-CHIP INTERCONNECT

In contrast to traditional methods, which make the usage of CNT-Cu ion dispersions, a novel manufacturing method for a CNT-Cu composite was developed. This method synergistically combines the benefits of both of the constituent materials. The prevailing process that is utilized for the production of Cu-CNT composites is discussed in the following subsections.

8.2.1 ELECTRODEPOSITION AND SYNTHESIS

Copper was electrodeposited within the pores of premade, macroscopic carbon nanotube (CNT) structures during this technique. Examples of these structures are buckypaper and CNT solids. On a substrate, large and vertically aligned single-wall CNT forests have been synthesized by opting for the water-assisted chemical vapor deposition approach. The forests had a diameter of ~3 nm, a height of ~500–700 μm, and a density of ~0.04 g cm^{-3}. The current techniques of synthesis have the drawback of producing impurities, which then needs to be eliminated by purifying stages that might cause the nanotubes to become damaged. As a direct result of interactions driven by van der Waals' forces, the fast accumulation of the smooth-sided tubes, either concurrent bundles or ropes are produced. This makes it difficult to disperse SWCNTs in solutions for further processing. Recently, the authors provide a logical, but generic synthesis strategy that is straightforward and overcomes all of these concerns at the same instant. In this technique, the activity as well as lifespan of the electrocatalyst are significantly increased with the insertion of a regulated quantity of water vapor in the growth conditions. This allows for more control over the amount of water vapor that is introduced. Researchers sought to explore a weak oxidizer that might precisely eliminate amorphous carbon but does not harm the nanotubes at the manufacturing temperature. This is due to the fact that covering the electrocatalyst agent using amorphous carbon atoms at the time of chemical vapor deposition (CVD) reduces their activity as well as lifespan. The research demonstrated that the presence of water stimulates and maintains the catalytic activity. The growth of SWCNTs was achieved using ethylene CVD by combining argon or helium with H$_2$ that included a limited and precisely regulated volume of water vapor. To enhance the catalytic lifespan, it was essential to strike a balance between the relative amounts of water and ethylene. The water-assisted synthesis method was successfully applied to a wide range of catalysts, namely Fe particles, that were used to produce SWCNT bundle [25]. Using FeCl$_3$ and sprayed metal (Fe, Al/Fe) on the silicon substrate, metallic foil, and the quartz illustrates the universal applicability of this technique.

8.2.2 LIQUID DENSIFICATION

The liquid densification process was used to convert the sparse forests into densely packed and aligned CNT structures. In order to produce high-density SWCNT bulk solid, this method makes use of a liquid-induced collapse of SWCNT bundle. Earlier study of densification of CNT particles had already demonstrated fascinating surface morphology constituted from vertically aligned MWCNT bundle or used to strengthen buckypaper or MWNT sheets. The majority of these studies employed relatively short MWCNT forests, ranging from 1 to 60 mm in length, which were still bound to the electrocatalyst. The length of the tube, its stiffness, and its surface adhesion all had a role in the formation of pyramid type topology and cellular configuration, which were perceived as a consequence. The utilization of lengthy (on a millimeter scale), vertically aligned SWCNTs that are extracted from the substrate as a single unit, consistently densified, and designed into basic forms is the most significant distinction brought about by this process [26]. Researchers normally collapsed as-grown forests by using a diluted ethyl alcohol mixture. In this method, the detached forests from the catalytic sides are put in a tiny beaker carrying the alcohol solvent, wherein collapse mechanism started. As the liquid penetrates the inter-tube zones, the size and shape of the forests reduce, and the forests sink. During that moment, researchers retrieve these from the solvent and let it to dry on a smooth silicon substrate, where the last step of the collapse happens.

8.2.3 Nucleation

In this method, the CNT has been converted in the form of a Cu/CNT hybrid structure using a process of nucleation growth electrodeposition as shown in Figure 8.4. This CNT-Cu hybrid was made possible through the development of a few critical processes. Initially, the process of electrodeposition has been divided into the following stages:

1. Introducing copper ions into an organic solvent and hydrating the hydrophobic CNTs in order to nucleate copper seeds on the carbon atoms.
2. Developing the Cu seeds in a liquid solution when all of the porous structure in the seeds are occupied [23].

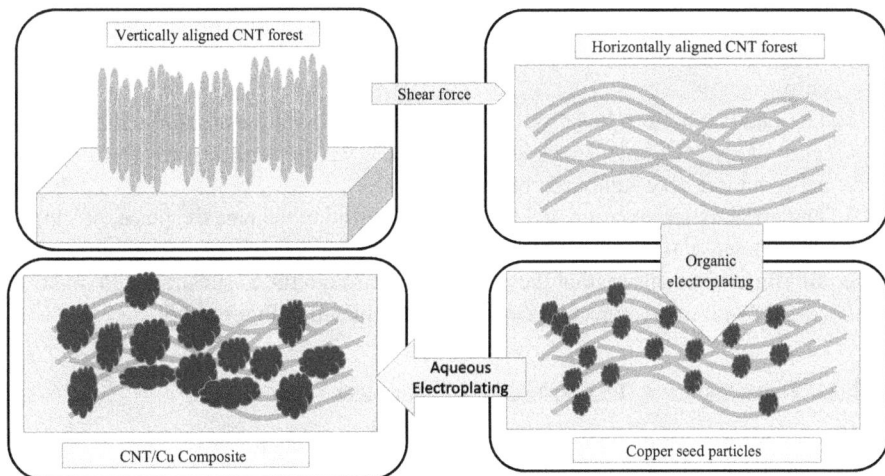

FIGURE 8.4   Fabrication steps of Cu-CNT composite horizontal interconnect.

Later, in the organic electrodeposition stage, the consistent nucleation precondition is that the rate-limiting stage be the copper seeding on the CNT structures instead of the Cu particle diffusion onto the CNT forest. This was because copper nucleation on CNTs is more stable than ion diffusion. In order to electrodeposit copper, all throughout the CNT forest and not only on the surface, gradual deposition rates (between 1 and 5 mA cm^{-2} in comparison to the normal rates of 50–100 mA cm^{-2}) [27] were necessary. At the aforementioned low current densities, Cu nucleated in a homogeneous manner across the CNT forest. This could result in good conductivity, a high Cu filling ratio in the Cu-CNT composite structure.

8.3 FABRICATION ASPECT OF HYBRID CU-CNT COMPOSITE-BASED VERTICAL INTERCONNECT

This section explains how Cu-CNT vertical aligned composites were created using the chemical vapor deposition process. A tape-assisted CNT transfer approach has been used in order to transport CNTs into a designated TSV aperture in a three-dimensional integrated circuit. It was performed to overcome the incompatibility issues that existed between the CNT growth and IC manufacturing operations. In the next steps, organic and aqueous electroplating solutions were applied to keep infiltrating the CNT forest. The widespread method that is used for the fabrication of Cu-CNT composites is detailed in the following subsections.

8.3.1 Synthesis of CNT Array Structures

In order to construct CNT-Cu composite TSVs that integrate the properties of both copper and carbon nanotubes, the design and manufacturing of a circular CNT forest with a diameter in the region of 200 μm is accomplished. During the electroplating process, holes deliberately formed in the CNT grid array that enabled copper to pass through. The particular steps are utilized in order to synthesize the CNT grid arrays as stated:

1. Initially, the circular array design is created using photolithography. Evaporation is used to deposit 10 nm Al$_2$O$_3$ and 1 nm iron layer onto the Si chip.
2. The catalyst pattern has produced into the Si substrate using typical lift-off technology, as shown in Figure 8.5a. The CVD is used to grow the CNTs using an industry standard equipment.
3. The growing temperature and time were adjusted to manage the place, height, and amount of the CNTs. The CNTs were grown to a height of several hundred micrometers to ensure that they could pass through the Si substrate and make contact with interconnect on both sides as depicted in Figures 8.5b and c.

8.3.2 Processing of TSV and CNT Placement

TSV holes have been etched onto a Si wafer utilizing deep reactive ion etching in concurrence with the CNT synthesis method as shown in Figure 8.5d. The target chip has been patterned applying standard lithography techniques before DRIE procedure. The vertical interconnect access pattern has also been employed to align with

(a)

(b)

(c)

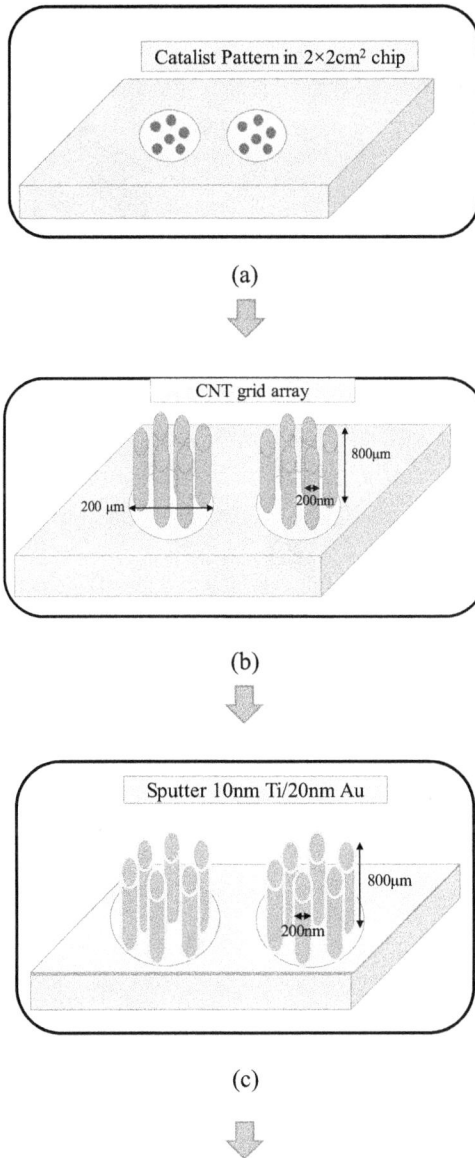

FIGURE 8.5 Schematic representation of the method followed in order to make composite Cu-CNT TSVs. (a) By using E-beam evaporation and the assistance of bilayer photoresists, patterned catalysts were successfully deposited. (b) A CVD process with a growth temperature of 700°C was used to fabricate a CNT grid array. (c) Sputtered 10 nm of titanium and 20 nm of gold inside the CNT forest. (d) and (e) Meanwhile, the target silicon substrate with TSV was created using deep reactive ion etching (DRIE). (f) The adhesive tape was left in place while thermal release tape was adhered to the front surface of the target chip. Following this, the CNT grid array was placed into the TSV utilizing a flip-chip bonder. (g) The donor wafer or chip was taken out of the process. (h) After the adhesive tape has been detached, the copper was electroplated and deposited into the vias to produce the composite CNT-Cu TSV.

(Continued)

(d)

(e)

(f)

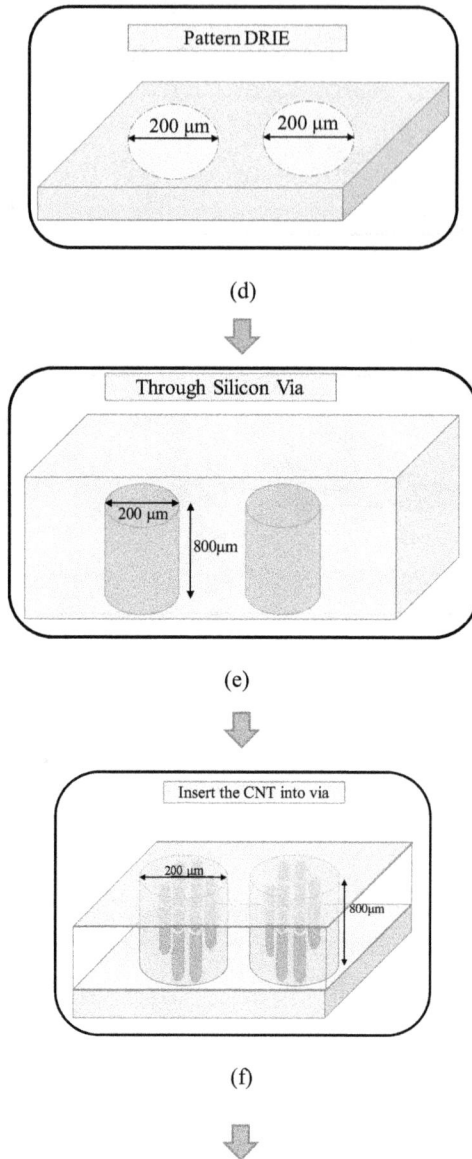

FIGURE 8.5 (*Continued*)   Schematic representation of the method followed in order to make composite Cu-CNT TSVs. (a) By using E-beam evaporation and the assistance of bilayer photoresists, patterned catalysts were successfully deposited. (b) A CVD process with a growth temperature of 700°C was used to fabricate a CNT grid array. (c) Sputtered 10 nm of titanium and 20 nm of gold inside the CNT forest. (d) and (e) Meanwhile, the target silicon substrate with TSV was created using deep reactive ion etching (DRIE). (f) The adhesive tape was left in place while thermal release tape was adhered to the front surface of the target chip. Following this, the CNT grid array was placed into the TSV utilizing a flip-chip bonder. (g) The donor wafer or chip was taken out of the process. (h) After the adhesive tape has been detached, the copper was electroplated and deposited into the vias to produce the composite CNT-Cu TSV.

(Continued)

(g)

(h)

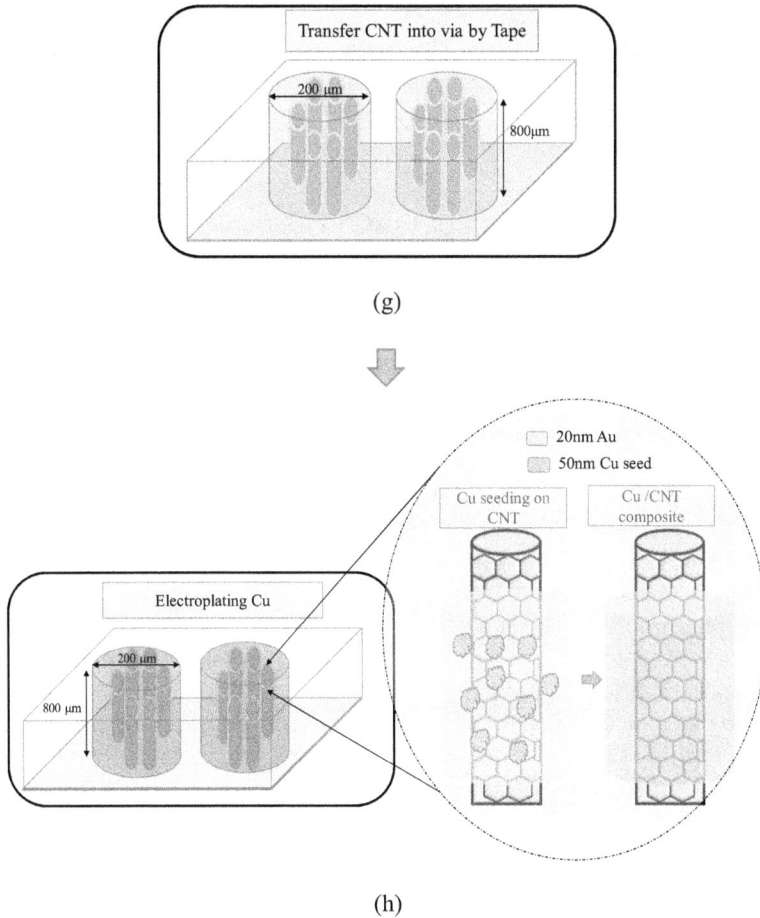

FIGURE 8.5 (*Continued*) Schematic representation of the method followed in order to make composite Cu-CNT TSVs. (a) By using E-beam evaporation and the assistance of bilayer photoresists, patterned catalysts were successfully deposited. (b) A CVD process with a growth temperature of 700°C was used to fabricate a CNT grid array. (c) Sputtered 10 nm of titanium and 20 nm of gold inside the CNT forest. (d) and (e) Meanwhile, the target silicon substrate with TSV was created using deep reactive ion etching (DRIE). (f) The adhesive tape was left in place while thermal release tape was adhered to the front surface of the target chip. Following this, the CNT grid array was placed into the TSV utilizing a flip-chip bonder. (g) The donor wafer or chip was taken out of the process. (h) After the adhesive tape has been detached, the copper was electroplated and deposited into the vias to produce the composite CNT-Cu TSV.

the pre-existing CNT cluster structures for subsequent chip and TSV transfers. The through silicon via configuration is depicted in Figure 8.5e. A robust transfer technique named as tape-assisted transfer is employed to introduce CNTs into TSV holes in terms of making the CNT growth parameters compatible with the TSV assembly process [28]. The target chip was initially taped using thermal release adhesive on the backside. The purpose of such tape would be to maintain CNTs within the via apertures while the donor wafer is being removed. Second, a flip-chip bonder has been used to adjust the CNTs to the TSV holes and firmly push the CNTs into the via holes such that the CNTs touched the tape. Finally, the donor chip was removed using the vacuum nozzle of the flip-chip bonder, with the CNTs retaining within the TSV holes owing to the tape's adhesiveness.

8.3.3 ELECTROPLATING METHOD OF CU-CNT HYBRID

Copper is plated into the CNT grid array by utilizing a two-stage electroplating process that results in a new composite material with a great electrical conductivity and mechanical performance [23]. Before beginning the electroplating process, the surface of the samples is sputtered with a titanium-gold sandwich layer that possesses a thickness of 10 nm titanium and 20 nm gold, as depicted in Figure 8.5c. The sandwich layer exhibits two purposes: first, it increases the CNTs' wetting capacity in the copper electroplating solution, and second, it serves like a conductive layer in the electroplating process. Both of these functions are accomplished by this layer. After that, the CNT arrays are left submerged in a Cu acetate electrolyte in acetonitrile with a dimension of 2.75 mm while being subjected to a Galvanostatic electric field in order to seed copper into the CNT arrays in a uniform manner. Thereafter, the sample is put through a second aqueous electrodeposition plating technique so that the copper seeds could be developed. In the first step, a current density that is much lower than normal (only 5 mA cm^{-2} as compared to the standard 30 mA cm^{-2}) is utilized so that the nucleation of copper ions would occur at a slower pace within the CNT grid array. In the second step of the plating process, performed with a standard current density of 30 mA cm^{-2}, the process ends.

8.4   ELECTRICAL MODELING OF CU-CNT COMPOSITE-BASED ON-CHIP INTERCONNECT

A hybrid structure of Cu-CNT-based horizontal interconnect above a ground plane, comprising of N identical CNTs arranged evenly inside the Cu wire, is depicted in Figure 8.6 and their associated physical parameters are summarized in Table 8.1. As shown in Figure 8.6b, the Cu/CNT composite interconnect is comprised of CNTs that could be single-walled or multi-walled. For MWCNT, the outermost diameter is designated as D_{out}, whereas the innermost is represented as D_{in}. In general, the innermost diameter of MWCNT can be considered to be half of its outermost diameter, or $D_{in} = D_{out}/2$ [29].

Furthermore, l_{comp}, w_{comp}, and t_{comp} represent the horizontal interconnect length, width, and thickness, respectively, while d_g represents the distance from the ground plane to the horizontal interconnect. The amount of CNT filled in the Cu/CNT hybrid structures represented in the form of CNT filling ratio would be calculated as follows:

$$f_{CNT} = \frac{N\pi\left(D_{out} + .31nm\right)^2}{4w_{comp}t_{comp}} \tag{8.1}$$

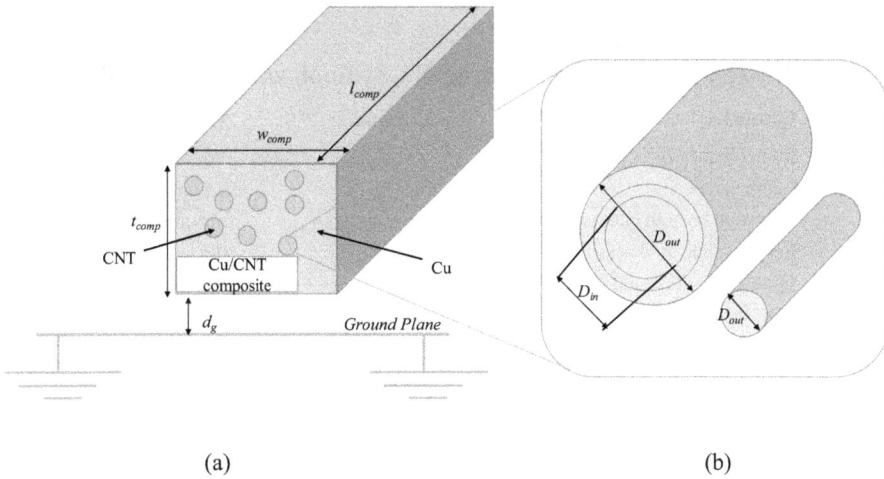

FIGURE 8.6 Horizontal interconnect above the ground plane: (a) Cu-CNT composite and (b) MW and SW CNT structure.

TABLE 8.1
Physical and Electrical Parameters, Symbol, and Its Value [5]

Parameter	Symbol	Values
Length of horizontal interconnect	l_{comp}	50 μm
Width of horizontal interconnect	w_{comp}	800 nm
Thickness horizontal interconnect	t_{comp}	900 nm
Outer diameter of SWCNT/MWCNT	D_{out}	3 nm
Inner diameter of MWCNT	D_{in}	1 nm
Conductivity of Cu	σ_{Cu}	5.8×10⁸ S/m
Conductivity of Si	σ_{si}	10 S/m
Conductivity of RDL	σ_{RDL}	5.8×10⁸ S/m
Conductivity of bump	σ_{bump}	9.1×10⁸ S/m
Relative permittivity of liner	ε_{r_liner}	3.9
Relative permittivity of Si	ε_{r_si}	11.68
Relative permittivity of oxide	ε_{r_ox}	3.9
Relative permittivity of IMD	ε_{r_IMD}	2.6
Relative permittivity of underfill	$\varepsilon_{r\_Underfill}$	7
Relative permittivity of passivation	$\varepsilon_{r\_passive}$	4.1
Relative permittivity of air	ε_{r_air}	1
Relative permittivity of dielectric	ε_{r_diel}	2.6
TSV height	h_{comp}	60 μm
TSV diameter	d_{comp}	30 μm
TSV pitch	P	100 μm
IMD height	h_{IMD}	7.1 μm
Liner thickness	t_{liner}	0.52 μm

(*Continued*)

TABLE 8.1 *(Continued)*
Physical and Electrical Parameters, Symbol, and Its Value [5]

Parameter	Symbol	Values
Bump diameter	d_{Bump}	70 μm
Bump height	h_{Bump}	15 μm
Depletion layer thickness	t_{dep}	0.75 μm
Bottom oxide thickness	t_{bot}	0.5 μm
Distance from ground plane	d_g	100 μm
RDL thickness	t_{R_comp}	5 μm
RDL length	l_{R_comp}	300 μm
RDL width	w_{R_comp}	100 μm
RDL spacing	s_{R_comp}	250 μm
Passivation height	h_{passiv_comp}	7 μm
Dielectric height	$h_{dielectric\_comp}$	7.1 μm
Substrate height	h_{sub_comp}	235.9 μm
Driver resistance	R_{driver}	16.67 kΩ
Driver capacitance	C_{driver}	0.049 fF
Load capacitance	C_{Load}	0.14 fF

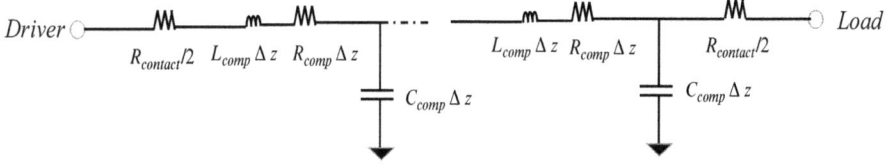

(a)

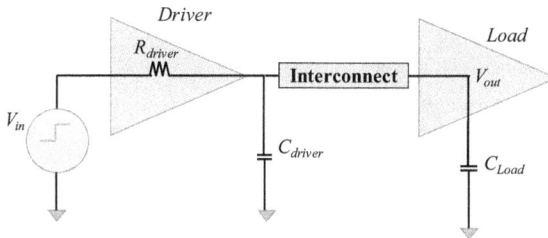

(b)

FIGURE 8.7 Driver interconnect and load model: (a) electrical model and (b) schematic.

The model takes into account a spacing of 0.155 nm among the CNT and Cu atoms [30]. The space between adjacent shells in the MWCNT is 0.34 nm, or the van der Waals' gap. Figure 8.7 shows the resistance, inductance, and capacitance (*RLC*) model for the driver-interconnect-load (DIL) arrangement.

The input and output voltages of the driver and load are represented as V_{in} and V_{out}, respectively, the driver electrical capacitance and resistance are C_{driver} and R_{driver}, respectively, and the load capacitance is C_{Load}. It is sufficient to perform transient analysis using the RC structure of MOS transistor. If a greater level of precision is desired, then short-channel effects may be described with the help of the alpha-power MOSFET design [31,32]. The contact resistance $R_{contact}$ is calculated as

$$R_{contact} = \frac{f_{CNT}}{N} \times \left(R_{mc} + \frac{R_Q}{N_{ch}} \right) \tag{8.2}$$

where R_{mc} and $R_Q = h/(2e^2) = 12.9$ kΩ denote the imperfect and quantum contact resistances, N_{ch}, N, h, and e represent the number of conducting channels of an individual CNT, total number of nanotubes, Planck's constant, and the electronic charges, respectively.

8.4.1 Modeling of RLGC Parameters

Earlier, the quantum conductance was calculated by employing the density of state (DOS); nevertheless, the quantum conductance has not been in any way reliant on the DOS. Although this method does not rely on the DOS, it recommends that there may be a further basic approach to derive quantum conductance from fundamental principles. The fundamental method to consider this problem is to begin with the simplest assumption of quantum mechanics as per Max Planck's theorem:

$$E = hf \tag{8.3}$$

where f represents the frequency of the interaction between particle and wave. It makes it even more fascinating to express the concept in the form of electron flux. It is due to the fact that researchers are more concerned with the quantity of electrons that pass through a given area in a predetermined period of time. Hence the afore-mentioned Eqn. (8.3) can be expressed as

$$E = hf = \frac{h \times 2e}{2e \times \tau_e} = \frac{h \times I_{Electrons}}{2e} \tag{8.4}$$

where τ_e represents the period of an electron and $I_{Electron}$ denotes the electron current while including the electron spin that possesses an energy E. The energy degeneration that takes place at energy E is not taken into account by this current. Hence, the current of electron can be expressed as

$$I_{Electrons} = \frac{2e \times E}{h} \tag{8.5}$$

The magnitude of the net current in a one-dimensional channel that is connected to reservoir contacts is determined by the variance in the chemical potentials of the reservoirs. If every state is inhabited up to μ_S at the source reservoir and μ_D at the drain reservoir, with $\mu_S - \mu_D = eV$ the difference between the two, the resulting value of the channel's net current I is represented as follows:

$$I = \frac{2e}{h}(\mu_s - \mu_D) = \frac{2e^2}{h}V \tag{8.6}$$

and the quantum conductance emerges logically as a consequence of this, which is presented as

$$G_0 = \frac{2e^2}{h} \tag{8.7}$$

In a similar way, the quantum capacitance of the carbon nanotube could likewise be determined in a simple way using the quantum conductance of the material. The time-varying current in a capacitor may be obtained using the displacement current relation that is given as follows:

$$I(t) = \frac{dQ(t)}{dt} = v_F \frac{dQ(t)}{dl} \tag{8.8}$$

wherein Q refers to the total charge that is stored in the capacitor, and dQ/dl represents the charge density, which is directly related to the product of the capacitance in a unit length and the time-variable voltage difference $V(t)$:

$$I(t) = v_F C_q V(t) \tag{8.9}$$

As a result, the quantum capacitance per unit length is

$$C_q = \frac{1}{v_F} \times \frac{I(t)}{V(t)} = \frac{G_0}{v_F} = \frac{2e^2}{h v_F} \tag{8.10}$$

It's worth noting that the concept of quantum capacitance for the metallic CNT could also be applied in the semiconductor CNT by substituting the v_F with the thermally averaged particle velocity. The quantum conductance may also be used to compute the kinetic inductance of metallic CNT:

$$V(t) = \frac{d\varphi(t)}{dt} = v_F \frac{d\varphi(t)}{dl} \tag{8.11}$$

where $d\varphi/dl$ represents the equivalent flux density that may be attributed to kinetic inductance. This can be computed by multiplying the current by the kinetic inductance per unit length.

$$V(t) = v_F L_k I(t) \tag{8.12}$$

As a result, per-unit-length kinetic inductance of the CNT can be expressed as

$$L_k = \frac{1}{v_F} \times \frac{V(t)}{I(t)} = \frac{1}{v_F G_0} = \frac{h}{2e^2 v_F} \tag{8.13}$$

8.4.2 Equivalent Cu-CNT Composite Model

The lumped electrical model parasitic R_{comp}, L_{comp}, and C_{comp} represent the per-unit-length (p.u.l.) interconnect resistance, inductance, and capacitance, respectively, as depicted in Figure 8.6. The Cu-CNT composite interconnect scattering resistance in p.u.l. is represented as

$$R_{comp} = \frac{\rho_{eff\_comp}}{w_{comp} \times t_{comp}} = Re\left(\frac{1}{\sigma_{eff\_comp}}\right) \times \frac{1}{w_{comp} \times t_{comp}} \tag{8.14}$$

$$\sigma_{eff\_comp} = (1 - f_{CNT})\sigma_{Cu} + f_{CNT}\sigma_{CNT} \tag{8.15}$$

where σ_{eff_comp} and $\rho_{eff\_comp} = Re(1/\sigma_{eff\_comp})$ are the overall complex conductivity and resistivity of the hybrid Cu-CNT structure, respectively; σ_{Cu} and σ_{CNT} denote the Cu conductivity and CNT conductivity, respectively. The σ_{CNT} can be calculated by using [33] as

$$\sigma_{CNT} \approx \frac{F_m \times 4l_{comp}}{Z_{CNT}\pi(D_{out} + .31nm)^2} \tag{8.16}$$

wherein F_m represents the metallic carbon nanotube fraction and Z_{CNT} denotes the isolated CNT self-impedance. In general, the metallic fraction $F_m = 1$ for multi-walled carbon tubes. The self-impedance (Z_{CNT}) in the case of single-walled or a particular shell in a multi-walled CNT can be given by

$$Z_{SWCNT}(D_{out}) = \frac{l_{comp}}{N_{ch}}\left(\frac{R_Q}{l_{mfp}} + sL_K\right) \tag{8.17}$$

For SWCNTs, $N_{ch} = 2$, whereas for MWCNTs, it depends on the temperature parameter [34]. The length of mean free path is denoted as $l_{mfp} = 1,000D_{out}$. The inductance of the Cu-CNT hybrid structure is comprised of both inner and outer inductances and can be expressed as

$$L_{comp} = L_{inner} + L_{outer} = Im\left(\frac{1}{\sigma_{eff\_comp} \times w_{comp} \times t_{comp}}\right) + \frac{\mu_0\varepsilon_0\varepsilon_r}{C_{E\_comp}} \tag{8.18}$$

where ε_0, ε_r, and μ_0 are the permittivity of vacuum and surrounding dielectric, permeability of vacuum, respectively, and the p.u.l. electrostatic capacitance C_{E_comp} can be given by [35]

$$C_{E\_comp} = \varepsilon_0\varepsilon_r\left\{\frac{w_{comp}}{d_g} + \frac{4}{\pi}\ln\left(1 + \frac{t_{comp}}{d_g}\right) + \frac{6}{\pi} + \frac{2}{\pi}\ln\left[1 + \frac{\pi w_{comp}}{2(1+\pi)(t_{comp} + d_g)}\right]\right\} \tag{8.19}$$

The Cu-CNT hybrid structure quantum capacitance is represented as

$$C_{Q\_comp} = NN_{ch} \times \frac{2e^2}{hv_F} \tag{8.20}$$

Since C_{Q_comp} is indeed higher than C_{E_comp}, the per-unit-length Cu-CNT composite interconnect capacitance can be approximated as C_{E_comp}.

8.5 ELECTRICAL MODELING OF CU-CNT COMPOSITE VIA

This electrical lumped architecture of a TSV that would be an analytical circuit structure is discussed in this section. Based on the lumped model, the *RLGC* model expressions are presented for this electrical structure. Due to the fact that the circuit structure is generated from the physical model, it thoroughly defines the physical relevance of each parasitic element by making use of the closed-form equations. The electrical model is presented with the structural attributes and material quality as variables in the analytical *RLGC* equations, so that it may be scaled appropriately. The construction of a TSV as well as its physical properties is shown in Figures 8.8 and 8.9. Within a single-ended signaling arrangement that makes use of the via-last approach, Figures 8.8 and 8.9 illustrate the construction as well as the features of a signal-ground (SG) TSV and bump. After completion of the metallization, a certain sort of TSV known as a via-last TSV is produced. As a direct consequence of this, a via-last TSV passes not only through the substrate but also through the layer known as the inter-metal dielectric (IMD). Due to a required connectivity of TSV between the top and the bottom dies, it must go through the silicon substrate in a vertical direction. The via that it passes through is then filled with metal to ensure proper electrical conduction. Copper, tungsten, and a Cu-CNT composite are examples of conductors that may be used as filler materials [1,7]. The Cu-CNT composite filler is the approach that is now used most often. Silicon is frequently used as the substrate material to support semiconductor devices; however, glass and other organic substrates are also being explored as substrate materials to increase insulating performance [36]. Due to the significantly reduced length of the connection between chips, TSV-based 3D IC has a significant advantage over 2D or 2.5D integration achieved by wire bonding or flip-chip packaging. This presents a significant strategic advantage for TSV-based 3D IC. In addition, due to the requirement of a noticeably lesser

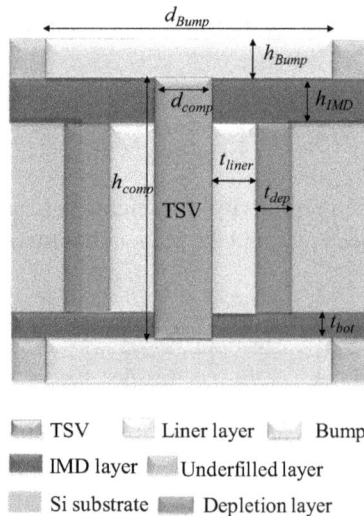

FIGURE 8.8 Physical structure of TSV.

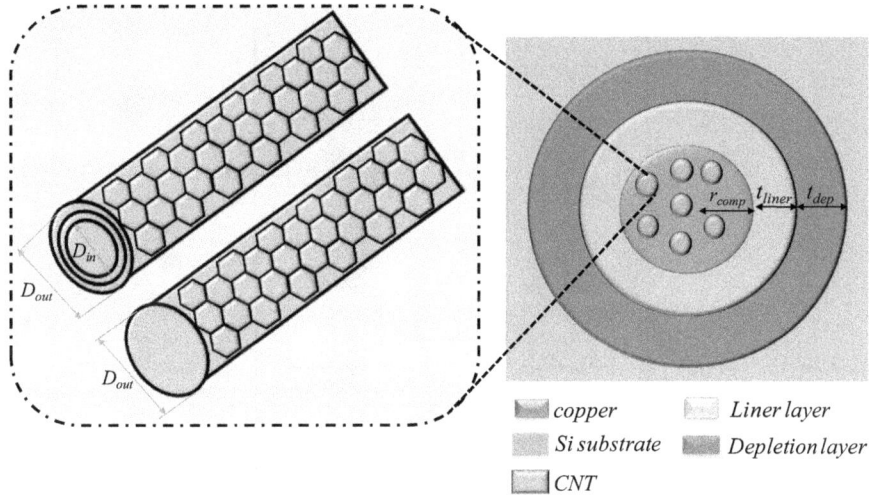

copper   Liner layer

Si substrate   Depletion layer

CNT

FIGURE 8.9 Top view of Cu-CNT-based TSV structure.

die area per I/O, the 3D integrated circuits based on TSV have a wide range of applications for I/O. The diameter to pitch ratio for TSV-to-TSV may range about 1:2.5 to 1:4, depending on the manufacturing procedure and the size of the keep-out zone that surrounds a TSV. The thin insulating layer that surrounds the TSV has a thickness ranging between 0.1 and 0.52 µm. The bottom oxide layer is another insulating layer that forms on the bottom side of the silicon substrate. This insulation layer has a thickness of 0.1–0.5 µm and is necessary to be placed between the conductive silicon substrate and the metallic bump in order to provide insulation between the two components.

A high-frequency, scalable lumped electrical model of TSVs with bumps is illustrated by utilizing the physical structure as shown in Figure 8.10. In order to formulate the analytical *RLGC* equations, the physical setup as well as the design parameters are utilized. As a consequence of this, each *RLGC* equation is a function of the variables that determine the design parameters. The total length of the TSV connection that includes both the TSV and the bump will eventually be reduced to tens of micrometers as processing technology advances. The lumped approximation is true, and the TSV model needs just a single lumped stage since its length is insufficiently tiny in relation to the wavelength of several or tens of GHz operating frequency. In order to accurately estimate the electrical behaviors of a TSV in the case of the via-first or via-middle process, the parasitic capacitances such as the capacitance of IMD layer (C_{IMD}) should either be ignored or other parasitic capacitances should be addressed in addition. This is because the capacitance of IMD layer can have a significant impact on the electrical behavior of a TSV. Additionally, if there are active circuits between TSVs, the expected field distribution can turn out differently than expected. A realistic design would often include the construction of a number of TSVs, including array-type TSVs, rather than a single ground-signal (GS) pair of TSVs. On the other hand, the equivalent circuit model of a TSV among

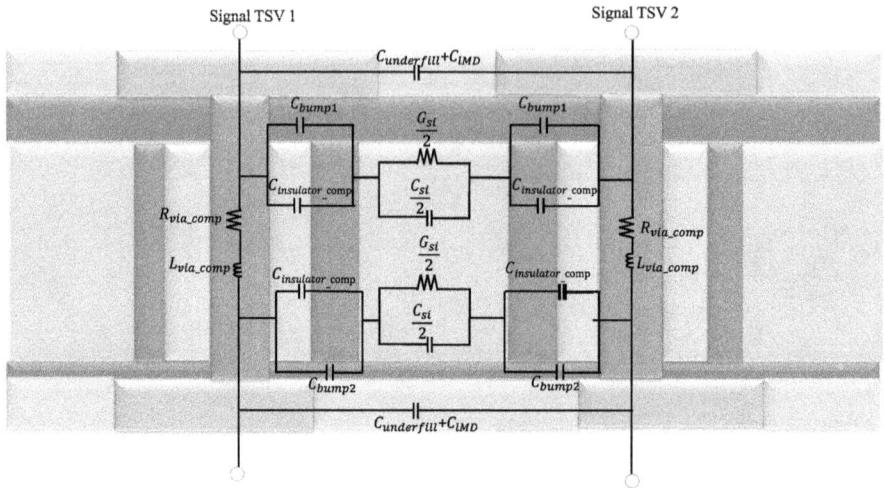

FIGURE 8.10 Electrical model of TSV.

the via array is the same as the proposed model represented in Figure 8.10, with the exception that different weights must be assigned to each *RLGC* component. As a consequence of this, one may attempt at TSV modeling of the array kind by either extracting or modeling the *RLGC* component weights.

Figure 8.8 displays a Cu-CNT hybrid structured TSV, implanted on Si substrate and defined by radius r_{comp} and height h_{comp}. Each Cu-CNT composite TSV contains N_{cnt} number of identical CNTs of dimension D_{out} that are equally dispersed in the Cu material, and the CNT filling ratio can be specified as

$$f_{cnt} = N_{cnt} \left(D_{out} + .31nm \right)^2 / 4r_{comp}^2 \qquad (8.21)$$

There should be a gap of 0.155 nm among the CNT and the copper, as stated in [30]. It is worth noting that the transverse transmissions that exist among the Cu and carbon atoms may be ignored since TSVs carry information in a vertical direction [37]. The complex conductivity of the hybrid structure could be determined as

$$\sigma_{eff\_comp} = (1 - f_{CNT})\sigma_{Cu} + f_{CNT}\sigma_{CNT} \qquad (8.22)$$

where σ_{CNT} stands for the effective complex conductivity of the CNT forest

$$\sigma_{CNT} = C_{CNT} f_{CNT} / Z_{CNT} \qquad (8.23)$$

where $C_{CNT} = \dfrac{4h_{comp}Fm}{\pi \left(D_{out} + .31nm \right)^2}$ and Z_{CNT} represents the inherent self-impedance of the individual carbon tubes. For SWCNT, the aforementioned impedance can be considered as [22]

$$Z_{CNT} = R_{mc} + \frac{h}{2q^2 N_{ch}} \left(1 + \frac{h_{comp}}{\lambda_{eff}} + j\omega \frac{h_{comp}}{2v_F} \right) \qquad (8.24)$$

where R_{mc}, h, q, λ_{eff} ($=1{,}000\,D_{out}$), and v_F ($\approx 8\times 10^5$ m/s) denote the imperfect contact resistance, the Planck's constant, electron charge, effective mean free path, and the Fermi velocity, respectively. N_{ch} indicates the number of conducting channels, which is equal to 2 for SWCNT case. It's worth noting that the imperfect contact resistance R_{mc} is strongly influenced by the fabrication process.

The metallic fraction ($F_m = 1$) for multiwall CNT primarily represents the metal-to-semiconductor ratio. MWCNT is the coaxial arrangement of SWCNT, and the self-impedance can be calculated by the parallel composition of each shall that can be obtained from Eqn. (8.24). The number of conducting channels of the shell in the MWCNT could be formulated as $N_{ch} = 0.0612D + 0.425$, wherein D denotes the diameter of the SWCNT or the shell of an MWCNT.

The impedance of Cu in the Cu-CNT composite can be computed by [38]

$$Z_{cu} = 2Z_{metal} + R_{sub} + j\omega L_{outer}$$

$$= \frac{2(1-j)J_0((1-j)r_{comp}/\delta_{cu})}{2\pi r_{comp}\delta_{cu}\sigma_{cu}J_1((1-j)r_{comp}/\delta_{cu})} + j\omega\frac{\mu}{\pi}\cosh^{-1}\left(\frac{P}{2r_{comp}}\right) \tag{8.25}$$

$$+ \frac{\alpha\mu}{2}\mathrm{Re}\left[H_0^{(2)}\left\{\left(\frac{1-j}{\delta_{si}}\right)(r_{comp} + t_{ox} + t_{dep})\right\} - H_0^{(2)}\left((1-j)P\Big/\delta_{si}\right)\right]$$

where δ_{cu} and δ_{si} denote the damping parameters and P represents the pitch (center to center distance) between the TSV.

$$\delta_{cu} = \sqrt{\frac{2}{\omega\mu\sigma_{cu}}} \tag{8.26}$$

$$\delta_{si} = \sqrt{\frac{2}{\omega\mu(\sigma_{si} + j\omega\varepsilon_{si})}} \tag{8.27}$$

The total Cu-CNT composite impedance using the parasitics shown in Figure 8.11 can be obtained as

$$Z_{comp} = \left(Z_{cu}^{-1} + N_{CNT}Z_{CNT}^{-1}\right)^{-1} \tag{8.28}$$

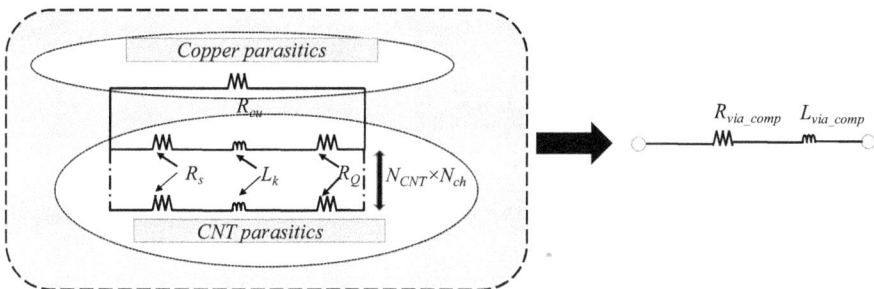

FIGURE 8.11   Electrical equivalent model of composite structure.

The total number of nanotubes in the Cu-CNT hybrid structure can be formulated by

$$N_{CNT} = \frac{\pi r_{comp}^2}{\left(D_{out} + .34nm\right)^2} \tag{8.29}$$

The total Cu-CNT composite resistance and inductance shown in Figure 8.11 can be expressed as

$$R_{via\_comp} = \text{Re}(Z_{comp}) + R_{bump\_comp} \tag{8.30}$$

$$L_{via\_comp} = \text{Im}(Z_{comp}) + L_{bump\_comp} \tag{8.31}$$

where the resistance and inductance of the bump can be represented by R_{bump_comp} and L_{bump_comp} that can be formulated by Eqn. (8.59), (8.72) and (8.73), respectively, in the following sections.

8.5.1 ELECTRICAL MODELING OF LINER, DEPLETION, BUMP, AND IMD LAYER

The insulating, i.e., oxide and depletion capacitances (C_{ox} and C_{dep}), could be formulated by imagining the structure as the concentric absolutely conducting cylinders having radii a and b separated using a perfect dielectric with permittivity ε_s. The positive z axis is positioned along the axis that is identical to all of the concentric cylinders as shown in Figure 8.12. The capacitance would be calculated by assuming that the inner wire has a total charge of Q_+ and integrating over the electric field that is associated with it to get the voltage that is present between the conductors. The resultant capacitance is then calculated by dividing the anticipated charge by the resultant potential difference.

$$C = \frac{Q_+}{V} \tag{8.32}$$

where $Q+$ denotes the charge associated on the positively charged conductor and V represents the potential measured between the negative and positive conductors.

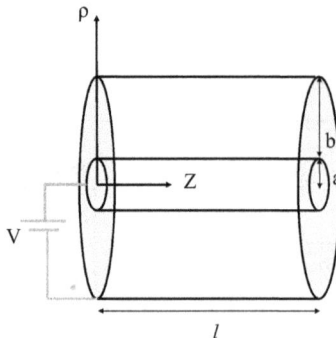

FIGURE 8.12   Capacitance model of coaxial structure.

The charge on the innermost conductor is evenly distributed with the density that can be expressed as

$$\rho_l = \frac{Q_+}{l} \tag{8.33}$$

The above expression (Eqn. 8.33) is measured in C/m. For calculating the electric field strength E, integrate the conductor along its path to obtain V, and then utilize Eqn. (8.32) to get the capacitance. Gauss' law is used to find the electric field induced due to a charged particle. This law necessitates integration across a surface enclosing the charge. As illustrated in Figure 8.13, considering a cylinder with a radius of a that is concentric with the z axis and is the most uniform with the charge distribution, it is anticipated to give the easiest possible solution. At first look, it seems to have a difficulty since the charge extends to infinity in the $+z$ and $-z$ directions, making it difficult to contain the whole charge. Therefore, by avoiding that problem and just selecting the finite length (l) of the cylinder, the issue for this finite-size cylinder that holds just a part of the charge can be solved. Later, with the assumption that the l tends to be infinity and considering the remainder of the charge through the cylinder, Gauss' law can be modified as

$$\oint_s D.ds = Q_{encl} \tag{8.34}$$

where $D=\varepsilon E$ represents the electric flux density, ds is the normal face of the closed surface S, and Q_{encl} denotes the enclosed charge.

The electric flux density as a function of charge q can be obtained as

$$D = \hat{r}\frac{q}{4\pi r^2} \tag{8.35}$$

By adding particles in pairs, an infinite line of charge can be constructed. In such case, a particle pair at a time is added, with the first particle located on the $+z$ axis and the second particle located on the $-z$ axis, both at the same distance from the origin. Consequently, other particle pairs can be incorporated in this manner until

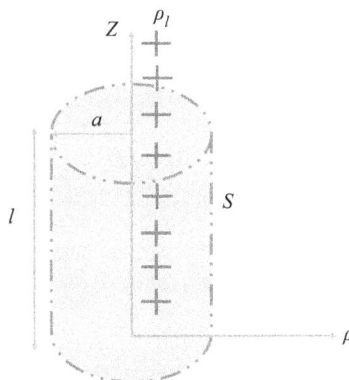

FIGURE 8.13 Line charge electric field.

the equivalent charge extends in both directions to infinity. According to the super-position principle, the resultant field will equal the total of the fields generated by the particles. As a result, D could not possess a factor in the $\hat{\phi}$ direction since none of the constituent particles' fields have a factor in that particular direction. Likewise, it can be demonstrated that the magnitude of D cannot be affected by ϕ since none of the component particles' fields are affected by ϕ and the charge distribution is the same with rotation in ϕ. Furthermore, the charge distribution above and below that plane of constant z is identical for any value of z; hence, D could be a factor of z and can-not have any constituent in the $\hat{z}$ direction. As a result, D must be directed radially outward, that is, in the $\hat{\rho}$ direction:

$$D = \hat{\rho}D_\rho(\rho) \tag{8.36}$$

The Q_{encl} on the RHS of Eqn. (8.34) equals $\rho_l l$ and can be expressed as

$$\oint_s \left[\hat{\rho}D_\rho(\rho) \right] . ds = \rho_l l \tag{8.37}$$

The cylinder S possesses flat top and both with the carved middle portion. Thus, Eqn. (8.37) can be expressed in the form of electric flux density as

$$\rho_l l = \int_{top} \left[\hat{\rho}D_\rho(\rho) \right]\left(+\hat{z}\, ds\right) +$$
$$\int_{side} \left[\hat{\rho}D_\rho(\rho) \right]\left(+\hat{\rho}ds\right) + \int_{bottom} \left[\hat{\rho}D_\rho(\rho) \right]\left(-\hat{z}\, ds\right) \tag{8.38}$$

The top- and bottom-side integration possess zero value according to the result of the dot products. In other words, since D is perpendicular to the top and bottom surfaces, the flux through them is zero.

$$\rho_l l = \int_{side} \left[D_\rho(\rho) \right] ds \tag{8.39}$$

The integration of the side surface is an open cylinder that may have the radius $\rho = a$, hence, $D\rho(\rho) = D\rho(a)$, where a is constant over this surface. As a result:

$$\rho_l l = \int_{side} \left[D_\rho(a) \right] ds = \left[D_\rho(a) \right] \int_{side} ds \tag{8.40}$$

The remaining integral is just the side surface area and is calculated by $2\pi a l$. Therefore, by solving $D\rho(a)$, one can get

$$D_\rho(a) = \frac{\rho_l l}{2\pi a l} = \frac{\rho_l}{2\pi a} \tag{8.41}$$

For $\rho = a$, the electric flux density D is represented as

$$D = \hat{\rho} D_\rho(\rho) = \hat{\rho}\frac{\rho_l}{2\pi\rho} \tag{8.42}$$

where the flux density $D = \varepsilon E$.

$$E = \hat{\rho}\frac{\rho_l}{2\pi\varepsilon\rho} \tag{8.43}$$

The electric field has been remarked in radially outward direction away from the line charge and hence its amplitude reduces as the distance from the line charge increases.
    Furthermore, the potential V can be formulated as

$$V = -\int_c E.dl \tag{8.44}$$

where C is any channel between the negative and the positive charge outer and inner conductor, respectively. This is the path that has a radial direction traverse with constant ϕ and z. Thus, the potential can further be rewritten as

$$V = -\int_{\rho=b}^{a}\left(\hat{\rho}\frac{\rho_l}{2\pi\varepsilon_s\rho}\right)(\hat{\rho}\,d\rho) = -\frac{\rho_l}{2\pi\varepsilon_s}\int_{\rho=b}^{a}\frac{d\rho}{\rho} = +\frac{\rho_l}{2\pi\varepsilon_s}\int_{\rho=a}^{b}\frac{d\rho}{\rho} = +\frac{\rho_l}{2\pi\varepsilon_s}\ln\left(\frac{b}{a}\right)$$

$$\tag{8.45}$$

The capacitance C in terms of charge and potential can be written as

$$C = \frac{Q_+}{V} = \frac{\rho_l l}{\left(\dfrac{\rho_l}{2\pi\varepsilon_s}\right)\ln\left(\dfrac{b}{a}\right)} \tag{8.46}$$

Thus, the equivalent capacitance for the coaxial cable can be represented as

$$C = \frac{2\pi\varepsilon_s l}{\ln(b/a)} \tag{8.47}$$

The above-mentioned formula is dimensionally valid, using units of F. Also, keep in mind that the expression is solely affected by materials and geometry. The formula wouldn't have been affected by charge or voltage, implying non-linear behavior. For the lumped-element transmission line model parameters, just divide by l to obtain per-unit-length capacitance as

$$C' = \frac{2\pi\varepsilon_s}{\ln(b/a)} \tag{8.48}$$

The $RLGC$ components of the electrical model of through silicon via can be computed by using analytical expression that depends on the material, physical properties, and frequency. The physical structure of a TSV is used to create the suggested $RLGC$ electrical circuit. The capacitance of a via is the most critical component that controls the overall parasitic of the circuit among the numerous TSV parasitics. An insulating layer encircling the TSV is required to electrically insulate it from the conductive silicon substrate. As seen in Figure 8.8, there is a liner capacitance, C_{liner_comp}, as a result of this liner layer. If the substrate is completely biased to ground, the depletion capacitance must be provided in series to C_{liner_comp}. Because of the metal-filled TSV and the conductive Si substrate, the insulating capacitance can be calculated using the coaxial cable structure with capacitive effect provided in Eqn. (8.48). The expression of the capacitance can be obtained from Poisson's equation. As shown in Figure 8.10, the C_{liner_comp} is a function of r_{comp}, h_{comp}, and t_{liner}. As a TSV's insulator capacitance is divided into two parallel parts, as seen in Figure 8.10, the expression of $C_{insulator\_comp}$ is stated as half of the overall insulating capacitance. Hence, the overall liner capacitance and the depletion capacitance can be expressed as

$$C_{liner\_comp} = \left\{ \frac{2\pi\varepsilon_0\varepsilon_{r,liner}}{\ln\left(\frac{r_{comp} + t_{liner}}{r_{comp}}\right)} \times \left(h_{comp} - h_{IMD}\right) \right\} [F] \qquad (8.49)$$

$$C_{depletion\_comp} = \left\{ \frac{2\pi\varepsilon_0\varepsilon_{r,si}}{\ln\left(\frac{r_{comp} + t_{liner} + t_{dep}}{r_{comp}}\right)} \times \left(h_{comp} - h_{IMD}\right) \right\} [F] \qquad (8.50)$$

The total insulating capacitance as a serial combination of liner and depletion can be computed as

$$C_{insulator\_comp} = \frac{\left(C_{liner\_comp}^{-1} + C_{depletion\_comp}^{-1}\right)^{-1}}{2} [F] \qquad (8.51)$$

Owing to the fact that TSVs have a 3D structure in the lossy Si substrate, the leakage is produced in the Si substrate whenever a ground via is located in close proximity to a signal via. The Si substrate leakage current implies the impedance of the channel between the signal and ground via. The $C_{insulator\_comp}$ has a significant effect on the insertion loss of a via that is caused by the Si substrate leakage channel. Consequently, the insertion loss of a TSV is increasing along with the size of the via physical parameter. As shown in Figure 8.8, the TSV diameter may be reduced in size and the TSV oxide thickness can be created thicker to further reduce the via capacitance.

Apart from this, the conductance that is caused by the loss tangent of the insulating layer is not taken into consideration in the model. It does not have an impact on the insertion loss of a via when the conductivity of the Si substrate is assumed to be 10 S/m. The loss of a TSV is no longer dominated by the silicon substrate conductance once the conductivity of the Si substrate drops to a low level, as it does with a high resistivity of the Si substrate. Then, the resistance of a via or the oxide liner loss caused by the loss tangent of the liner layer dominates, which has an effect on the insertion loss of a TSV; accordingly, dielectric loss terms cannot be neglected. For the purpose of accurate high-frequency modeling of a through-glass via, it is necessary, for instance, to take into account the dielectric loss of the insulating layer when the substrate material is glass (TGV). In addition, an increase in temperature causes a reduction in the relative permittivity of the dielectric layer, which in turn leads to a reduction in the parasitic capacitance.

As shown in Figure 8.10, the capacitance across bump-to-silicon on the top of substrate, designated as C_{bump1}, is added to the capacitance of the insulator for the recommended equivalent circuit model to be accurate. This capacitance may be analytically obtained using parallel plate capacitor model. As depicted in Figure 8.10, the origin of electric field lines is located at the point where the bump and Si substrate create parallel plate using IMD as a dielectric layer. As a consequence of this, the value of C_{bump1} is directly related to the area while being inversely proportional to the inter-metal dielectric layer height, i.e., h_{IMD}. As a direct result, it is possible to determine C_{bump1} by using d_{comp}, d_{Bump}, and h_{IMD} and can be expressed as:

$$C_{bump1} = \frac{\pi \times \varepsilon_0 \varepsilon_{r,IMD}}{h_{IMD}} \times \left\{ \left(\frac{d_{Bump}}{2} \right)^2 - \left(\frac{d_{comp}}{2} + t_{liner} \right)^2 \right\} [\text{F}] \qquad (8.52)$$

Similarly, the bottom-side bump to Si substrate capacitance (C_{bump2}) needs to be added to $C_{insulator\_comp}$. The bottom oxide layer creates the aforementioned capacitance between Si substrate and the bottom-side bump. After the back grinding of the *Si* chip, an oxidation process results in the formation of this bottom oxide layer. This operation is essential since it exposes the TSV and allows it to be connected to another die. The C_{bump2} is further generated from the model of the parallel plate capacitor. As a result, the value of C_{bump2} is determined by d_{comp}, d_{Bump}, h_{Bump}, and the bottom oxide thickness (t_{bot}) as represented in Eqn. (8.53). Now, it can be observed that, if the diameter of the bump possesses a larger value, C_{bump2} also increases. Since the bottom oxide thickness is in the range of 0.1–0.5 μm, the impact of C_{bump2} significantly impacts as that of $C_{insulator\_comp}$ counterparts, if the diameter of the solder bump is larger than the via diameter. Fine-pitch TSVs are chosen as the TSV technology progresses and it can primarily minimize the via capacitive parasitic and chip area utilization for the large number of I/Os. Additionally, in order to reduce the via capacitance effectively, it is necessary to lower not only the diameter of the TSV but also the diameter of the bump. If a signal or a power transmission circuit is implemented using metal layers on both sides of the silicon substrate, it is possible to eliminate this capacitance from the *RLGC* circuit of the signal and ground TSV.

$$C_{bump2} = \frac{\pi \times \varepsilon_0 \varepsilon_{r,ox}}{t_{bot}} \times \left\{ \left(\frac{d_{Bump}}{2} \right)^2 - \left(\frac{d_{comp}}{2} + t_{bot} \right)^2 \right\} [\text{F}] \qquad (8.53)$$

Using the via-last fabrication process, the capacitance between the bump pair ($C_{underfill}$) and the signal-ground pair using IMD (C_{IMD}) has been considered for parasitic modeling. The portions of the TSV and bump that make its cross-section circular can be depicted in Figure 8.8. Owing to this fact, the model of the parallel wire capacitance may be used to obtain $C_{underfill}$ and C_{IMD} [39]. The aforementioned capacitances can be analyzed by the via pitch distance, denoted as P, the diameter of the via and bump, signified as d_{comp} and d_{Bump}, and the underfill and IMD layer height, symbolized as h_{Bump} and h_{IMD}, respectively. In order to simplify the computation, the inverse hyper cosine may be replaced with the natural logarithm when P/d_{comp} is greater than one. It implies that d_{comp} is relatively less compared to the pitch distance. Additionally, the via in 3D packaging system is denser and possesses fine pitch, wherein it can be presumed that the ratio of P and d_{comp} is less than 10. As a result, $C_{underfill}$ and C_{IMD} can be modeled as

$$C_{underfill} = \frac{\pi \times \varepsilon_0 \varepsilon_{r,Underfill}}{\cosh^{-1}\left(\dfrac{P}{d_{Bump}}\right)} \times h_{Bump} [F] \tag{8.54}$$

$$C_{IMD} = \frac{\pi \times \varepsilon_0 \varepsilon_{r,IMD}}{\cosh^{-1}\left(\dfrac{P}{d_{comp}}\right)} \times h_{IMD} [F] \tag{8.55}$$

The lateral side of the bump also contributes further parasitic capacitance to the circuit. The parasitic capacitance is created when the lateral side of the solder micro bump is sandwiched between the Si chip. The $C_{insulator\_comp}$ component of the proposed model needs to have it included. It is possible to compute using the conformal mapping technique applied to the vertical plate. Using this technique, a three-dimensional connection may be converted into a parallel plate structure. This capacitance is about in fF range as per the current technology node of TSV design. Hence, due to its smaller value, this fringing parasitic capacitance is not included in parallel with $C_{insulator\_comp}$ in the model that has been presented.

Therefore, the total TSV capacitance is given by

$$C_{insulator\_comp} = \frac{\left(C_{liner\_comp}^{-1} + C_{depletion\_comp}^{-1}\right)^{-1}}{2} \tag{8.56}$$

Moreover, the semiconducting nature of the Si produces the capacitance and conductance that can be obtained using the parallel wire model as given by [19]

$$C_{si} = \frac{\pi \times \varepsilon_0 \varepsilon_{r,si}}{\cosh^{-1}\left(\dfrac{P}{d_{comp}}\right)} \times \left(h_{comp} - h_{IMD}\right) [F] \tag{8.57}$$

$$G_{si} = \frac{\pi \sigma_{si}}{\cosh^{-1}\left(\dfrac{P}{d_{comp}}\right)} \times \left(h_{comp} - h_{IMD}\right) [F] \tag{8.58}$$

The bump resistance (R_{bump_comp}) is likewise modeled using structural parameters, as illustrated in Figure 8.8. The generation of an eddy current, also known as the "skin effect," causes high-frequency current to flow near to the surface of the current carrying conductor [40]. In order to compute R_{bump_comp} at a high-frequency range, the depth of penetration, known as skin depth, must be considered in order to model the resistance of bump. The term "skin depth" is determined by material parameters, its conductivity, as well as frequency. It results in an analytical equation of R_{bump_comp} that can be represented as

$$R_{bump\_comp} = \sqrt{\left\{\rho_{bump} \times \frac{h_{bump}}{\pi \times (d_{Bump}/2)^2}\right\}^2 + \left\{k_p\left(\rho_{bump} \times \frac{h_{bump}}{2 \times \pi \times d_{Bump}/2 \times \delta_{sd,bump} - \pi\delta^2_{sd,bump}}\right)\right\}^2} \tag{8.59}$$

$$\delta_{sd,bump} = \frac{1}{\sqrt{\pi f \mu_{bump}\sigma_{bump}}} \tag{8.60}$$

Furthermore, due to the "proximity effect", there is an increase in current on the conducting wire surface but it is not equally distributed nearby its perimeter and attracted to most of the current at the inside-facing surfaces of the conducting wires. This effect appears after the skin effect starts appearing on the given frequency range. As the signal-ground via pairs are arranged in closer proximity, the proximity factor k_p rises. The amplitude of the proximity factor, k_p, for TSVs and bumps is analyzed by the ratio P/d_{comp} or P/d_{Bump} [41].

Utilizing the concept of the two-wire transmission line, the equivalent bump inductance can be represented as

$$L_{bump\_comp} = \frac{1}{2}\left\{\frac{\mu_0\mu_{r,bump}}{2\pi} \times h_{bump} \times \ln\left(\frac{P}{d_{Bump}/2}\right)\right\}[H] \tag{8.61}$$

The inductance of the bump is analyzed by using the loop inductance modeling of the parallel conducting via pair. L_{bump_comp} is computed by utilizing structural characteristics of the solder bump, i.e., d_{Bump} and h_{Bump}, as well as its material properties, i.e., $\mu_{r,bump}$.

8.5.2 PARALLEL WIRE TRANSMISSION LINE INDUCTIVE EFFECT MODEL

At higher frequency range, the inductive effect of the conducting via dominates over its resistive impact. The in-depth modeling of the inductive effect can be considered by using a widely spaced two-wire transmission line with two wires of equivalent radii (r_w) transmitting equal and opposite directions of the current and placed by a distance of s, as depicted in Figure 8.14a. The conducting wires are assumed to be infinitesimally long (or exceptionally larger than their radii) that results in a negligible fringing field capacitance at the end of the conducting wires. Although the wires are widely placed, the current is following evenly throughout the wire cross-sections, as depicted in Figure 8.14a. As a result, the magnetic fields (B) of both the wire pairs create circles that are concentrated at the center of the wire. Hence, the magnetic flux density of the wire can be expressed as

$$B_\varphi = \frac{\mu_0 I}{2\pi r} \tag{8.62}$$

(a)

(b)

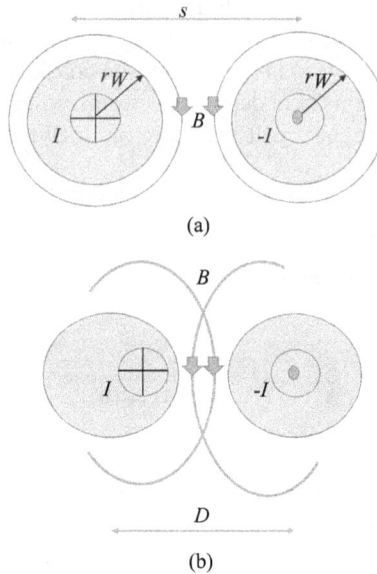

FIGURE 8.14   Parallel wire model with proximity effect: (a) far placed and (b) near placed.

where r represents the distance computed from the center of both parallel wire pairs. The equivalent flux across the parallel wire can be represented in the form of the current flow across each conducting surface (I), its material property, i.e., relative permeability (μ_0), and the distance computed from its center (r). Hence, the net equivalent flux can be computed as

$$\psi = 2 \int_{r_w}^{s-r_w} \frac{\mu_0 I}{2\pi r} dr = \frac{\mu_0 I}{\pi} \ln \frac{s-r_w}{r_w} \cong \frac{\mu_0 I}{\pi} \ln \frac{s}{r_w} \qquad (8.63)$$

Due to the widely separated pairs of the wires, one can assume that $s - r_w \cong s$. As a result, the per unit length of the uniformly distributed current carrying wire can be expressed as

$$l_{approximate} = \frac{\mu_0}{\pi} \ln \frac{s}{r_w} \text{ H/m} \qquad (8.64)$$

The currents will be focused toward the facing sides if the wires are placed closely as shown in Figure 8.14b, and the result of the inductance expression in (8.64) is an estimate since it is dependent on the uniformly distributed current across the wire. Hence, the inductance in case of the closely spaced wire $D \le s$ can be determined by

$$l_{exact} = \frac{\mu_0}{\pi} \ln \left[\frac{s}{2r_w} + \sqrt{\left(\frac{s}{2r_w}\right)^2 - 1} \right] \text{ H/m} \qquad (8.65)$$

It is noted that Eqn. (8.65) can be rewritten as (8.64) if $s \gg 2r_w$. The expression of the two-wire transmission line may be simplified to include the situation of a one-wire transmission line placed at a height h above a ground plane, as illustrated in Figure 8.15a.

(a)

(b)

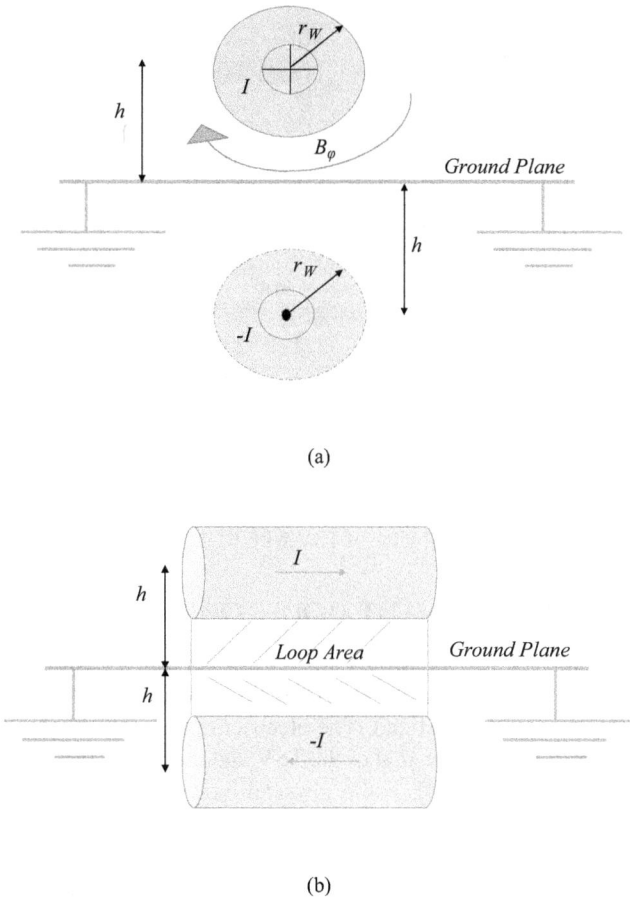

FIGURE 8.15 Transmission line model of single wire placed above the ground plane: (a) top view and (b) side view.

The main concept is to utilize the image approach to substitute the ground plane with the current image, as presented in Figure 8.15b.

Now, one can assume an analogous situation of a parallel wire line with a spacing of $s = 2h$. The loop area of one wire above a ground plane is between the wire surface and the ground, but the corresponding surface for the image method is between the wire surface and its image. As a result, the p.u.l. inductance considering the placement of one wire above the ground is one-half and that of a two-wire case, s, can be substituted by $s = 2h$. Consequently, the p.u.l. inductance of a transmission line with one wire above the ground is obtained as

$$l_{approximate} = \frac{\mu_0}{2\pi} \ln \frac{2h}{r_w} \text{ H/m} \tag{8.66}$$

$$l_{exact} = \frac{\mu_0}{2\pi} \ln \left[\frac{h}{r_w} + \sqrt{\left(\frac{h}{r_w} \right)^2 - 1} \right] \text{ H/m} \tag{8.67}$$

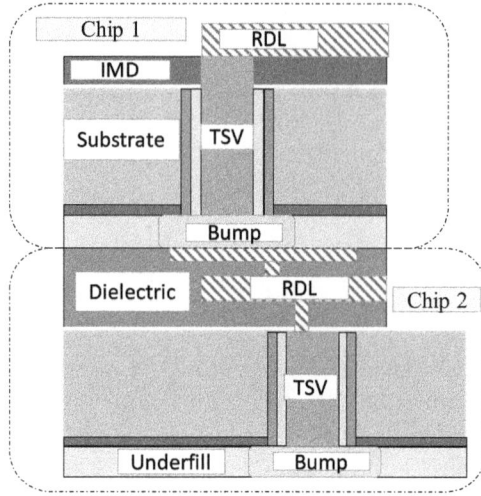

FIGURE 8.16   The cross-sectional view of TSV with RDL.

8.6   ELECTRICAL EQUIVALENT MODEL OF REDISTRIBUTION LAYER

Redistribution layer (RDL) is a metal connector that offers horizontal linkages between the chips of varying sizes. For the power distribution network of the via-last fabricated TSVs, the RDL connector is utilized. The electrical parameter of the RDL layer is modeled in the similar manner as that of the TSV and the bump. Figures 8.16 and 8.17 depict a single-ended signal RDL and a physical model of the RDL pair. However, Figures 8.18 and 8.19 depict the suggested equivalent electrical circuit model.

The RDL resistance can be obtained using the physical parameter shown in Figure 8.16. The current transmits across the RDL network as frequency rises, the frequency dependency is taken into account by computing the skin depth [40]. As a result, by including the skin effect, the resistance across the RDL can be expressed as

$$R_{RDL\_comp} = \sqrt{R_{dc,RDL\_comp}^{2} + R_{ac,RDL\_comp}^{2}} \ [\Omega/m] \tag{8.68}$$

where the dc and the ac components of the resistance can be given by

$$R_{dc,RDL\_comp} = \frac{\rho_{RDL}}{w_{R\_comp}t_{R\_comp}} [\Omega/m] \tag{8.69}$$

$$R_{ac,RDL\_comp} = \frac{\rho_{RDL}}{w_{R\_comp}\delta_{sd,RDL}} [\Omega/m] \tag{8.70}$$

where the skin depth of the RDL can be represented as

$$\delta_{sd,RDL} = \frac{1}{\sqrt{\pi f \mu_{RDL}\sigma_{RDL}}} [m] \tag{8.71}$$

It is worth noting that the resistance across the RDL is computed by presuming that the current distribution is focused on the bottom edge of the RDL. It is primarily generated by the electric field lines between the signal RDL and the Si substrate drawing the charge to the bottom edge. The current distribution focuses toward the lateral edge of the RDLs, if the gap between the parallel signal-ground RDLs is lower than the height between the signal RDL and the Si substrate. As a result, the AC component of the RDL resistance is differently modeled for the t_{R_comp}, and not for w_{R_comp}.

The high-frequency inductance of the RDL layer can be expressed by using the parallel wire transmission line that can be expressed by [42]

$$L_{RDL\_comp} = \frac{1}{2} \left\{ \begin{array}{l} \dfrac{\mu_0 \mu_{r,RDL}}{2\pi} \times \left[\ln\left(\dfrac{2l_{R\_comp}}{t_{R\_comp}} \right) - \dfrac{3}{4} \right] + \dfrac{\mu_0 \mu_{r,RDL}}{2\pi} \times \left[\ln\left(\dfrac{2l_{R\_comp}}{t_{R\_comp}} \right) - \dfrac{3}{4} \right] \\[2ex] -2 \dfrac{\mu_0 \mu_{r,RDL}}{2\pi} \times \left[\ln\left(\dfrac{2l_{R\_comp}}{S_{R\_comp}} \right) - 1 \right] \end{array} \right\} \text{[H/m]}$$

(8.72)

$$L_{RDL\_comp} = \frac{1}{2} \left\{ \frac{\mu_0 \mu_{r,RDL}}{2\pi} \times \left[2 \times \ln\left(\frac{S_{R\_comp}}{t_{R\_comp}} \right) + \frac{1}{2} \right] \right\} \text{[H/m]}$$ (8.73)

The structure of RDL is comparable to that of the on-chip metal model that is created in the IMD layer on Si substrate. As a result, the RDL model may be realized while using the model of the on-chip metal connection. The fringe capacitances between the RDL of the air, passivation, and dielectric layer must be considered to analyze the parasitic capacitances of the RDL, as illustrated in Figure 8.17. To represent the fringe capacitances between RDLs, the conformal mapping approach is utilized [43]. By using the aforementioned conformal approach, the RDL capacitance is formed as

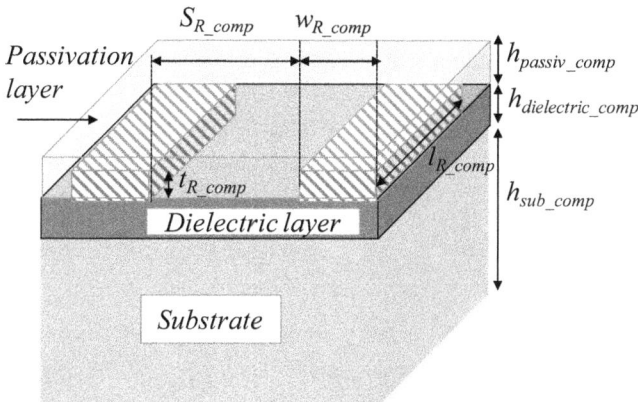

FIGURE 8.17 Physical parameter of the signal-ground RDL.

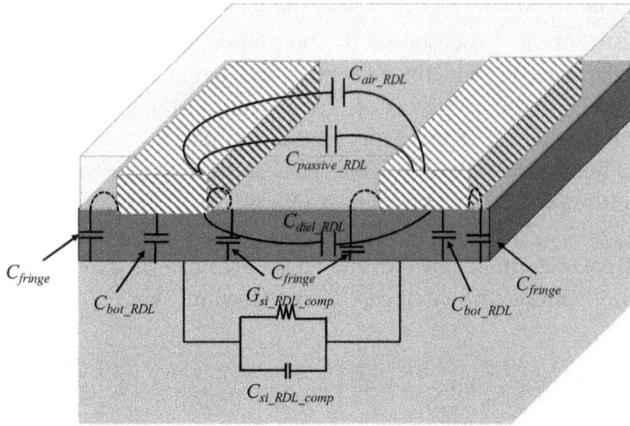

FIGURE 8.18   Internal fringing and capacitive model of RDL.

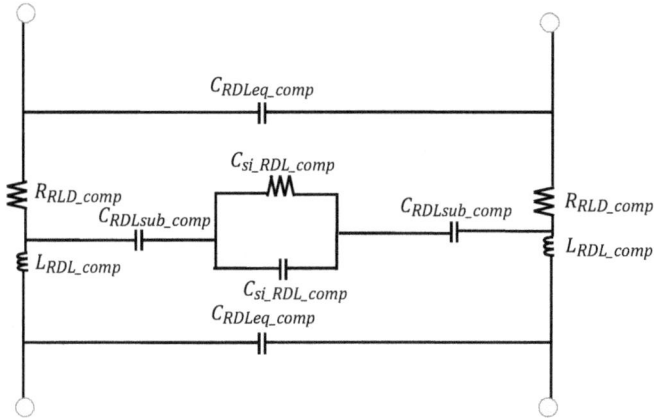

FIGURE 8.19   Equivalent $RLGC$ model of RDL.

the parallel combination of the air, passivation, and the dielectric capacitance. Hence, the equivalent RDL capacitance can be expressed as

$$C_{RDL\_to\_RDL} = C_{air\_RDL} + C_{passive\_RDL} + C_{diel\_RDL} \tag{8.74}$$

where the RDL capacitance of the air, passivation, and the dielectric is represented as

$$C_{air\_RDL} = \varepsilon_0 \varepsilon_{r,air} \frac{K'(k_0)}{K(k_0)} [\text{F/m}] \tag{8.75}$$

$$C_{passive\_RDL} = \varepsilon_0 (\varepsilon_{r,passive} - \varepsilon_{r,air}) \frac{K'(k_1)}{K(k_1)} [\text{F/m}] \tag{8.76}$$

$$C_{diel\_RDL} = \varepsilon_0 (\varepsilon_{r,diel} - \varepsilon_{r,passive}) \frac{K'(k_2)}{K(k_2)} [\text{F/m}] \tag{8.77}$$

where K is the first-kind elliptical integral.

Due to the electromagnetic field lines between the RDL and the Si substrate, the capacitance is formed and that capacitance between the RDL and the Si substrate is illustrated in Figure 8.17. Using the conformal mapping approach, the aforementioned capacitance can be modeled and computed as [44]

$$C_{RDL\_to\_sub} = C_{bot\_RDL} + 2C_{fringe}$$

$$= \varepsilon_0 \varepsilon_{r,SiO_2} \times \frac{w_{R\_comp}}{h_{dielectric\_comp}} + \varepsilon_0 \varepsilon_{r,SiO_2} \times \frac{K(k_{[VP]})}{K'(k_{[VP]})} \text{ [F/m]} \tag{8.78}$$

where

$$k_{[VP]} = \sqrt{1 - \left(\frac{h_{diel\_comp}}{h_{diel\_comp} + t_{R\_comp}}\right)^2} \tag{8.79}$$

The penetration of the electric field at the Si substrate between the signal and the ground RDL can create the capacitance and the conductance in the Si substrate that can be depicted in Figures 8.17 and 8.18. The capacitance and conductance as a function of its material ($\varepsilon_{r,eff}$) and the physical property (w_{R_comp}, h_{eff} [45]) can be expressed as Eqns. (8.80) and (8.81):

$$C_{si\_RDL\_comp} = \frac{\varepsilon_0 \varepsilon_{r,eff} w_{R\_comp}}{h_{eff}} \text{ [F/m]} \tag{8.80}$$

$$G_{si\_RDL\_comp} = \frac{\sigma_{eff} w_{R\_comp}}{h_{eff}} \text{ [F/m]} \tag{8.81}$$

where the effective relative permittivity and the conductivity can be expressed as

$$\varepsilon_{r,eff} = \frac{\varepsilon_{r,si} + 1}{2} + \frac{\varepsilon_{r,si} - 1}{2\sqrt{1 + 10\left(\dfrac{h}{w_{R\_comp}}\right)}} \tag{8.82}$$

$$\sigma_{eff} = \frac{\sigma_{si}}{2} + \frac{\sigma_{si}}{2\sqrt{1 + 10\left(\dfrac{h}{w_{R\_comp}}\right)}} \tag{8.83}$$

where the h_{eff} and h can be presented as

$$h_{eff} = \frac{w_{R\_comp}}{2\pi} \ln\left(\frac{8h}{w_{R\_comp}} + \frac{w_{R\_comp}}{4h}\right) \tag{8.84}$$

$$h = h_{dielectric\_comp} + h_{sub\_comp} \tag{8.85}$$

TABLE 8.2

Performance Comparison of Different Interconnect Material with Respect to Cu-CNT Composite

Interconnect Material	Improvement in Crosstalk Noise, Delay, and Power for Cu-CNT Composite in Comparison to		
	Peak Noise [46,47]	Crosstalk Delay [46,48]	Power Dissipation [47,48]
Cu	1.68×	2.26×	3.28×
SWCNT	1.50×	2.07×	3.14×
MWCNT	–	1.69×	3.05×

8.7 PERFORMANCE COMPARISON OF INTERCONNECT MATERIALS

With the development of several emerging interconnect material over the years, it is necessary to investigate their unique benefits and drawbacks and offer a comprehensive evaluation. The comparison assists in determining the optimum interconnect material that depends on individual needs. It also establishes the ground for emerging interconnect material approaches to be developed by acknowledging the inadequacies of existing material. Table 8.2 compares Cu, SWCNT, MWCNT, and Cu-CNT composite for several aspects such as peak noise, crosstalk delay, and power dissipation. The aforementioned parameters exhibit a degraded performance using individual Cu and CNT interconnect, whereas Cu-CNT composite interconnect possesses no such issue. It is due to the fact that the coupling ratio (C.R.) is reliant on coupling capacitance, and the electromigration resistance is the dominating factor for the peak-crosstalk-noise voltage and power dissipation. The remarkable electromigration resistance of the Cu-CNT composite and the reduced coupling capacitance result in the least delay and power dissipation. Hence, the Cu-CNT composite has the potential to strike a decent trade-off between the performance and the power analysis.

In addition, the term "Noise-Delay-Product" (NDP) refers to the end result of multiplying the crosstalk delay by the noise-peak voltage. It is a measurement that reflects how effectively the interconnect line maintains its signal integrity. The tendency is comparable to delay since it is the dominant element in NDP. The NDP of the Cu-CNT composite is reduced by 74% compared to the Cu counterparts. Hence, in terms of maintaining signal integrity, the Cu-CNT composite interconnect structure is superior to any other alternative filler material, making it an ideal solution for future VLSI interconnects.

SUMMARY

This chapter discusses the high-frequency *RLGC* model of a Cu-CNT composite-based through silicon via (TSV). The electrical model comprises not only the via but also the micro bump and the RDL layer, which are required when employing the TSVs for 3D IC packaging system. Analytic expressions generated from the physical parameter are used to create the electrical model. The parasitic equations have been modeled using the parallel wire transmission line and the conformal mapping method is dependent on the physical property of the TSV, bump, and the RDL layer. In addition, fringing effect of air, passivation, and the dielectric is considered for modeling of RDL layer. The RDL is positioned above a conductive silicon substrate that subsequently creates a functional electrical path. Therefore, it is essential to model the capacitance of the substrate as well as its resistance by utilizing the effective penetration depth of the electric field into the silicon substrate. Additionally, a comprehensive step-by-step explanation of the process of fabricating a TSV and horizontal interconnect that includes etching, electrodeposition and synthesis, liquid densification has been included. The Cu-CNT-based interconnect filler material plays a crucial role in the IC packaging system. Therefore, the comparative electrical properties of the different filler materials are also highlighted. Based on the comparative results of the different filler materials, the remarkable properties of the Cu-CNT composite are the main reason for its usage in the IC packaging system.

REFERENCES

[1] Kaushik, B. K. and Majumder, M. K. 2015. *Carbon Nanotube Based VLSI Interconnects Analysis and Design*. India: Springer.

[2] Majumder, M. K., Kumari, A., Kaushik, B. K. and Manhas, S. K. 2014. Analysis of crosstalk delay using mixed CNT bundle based through silicon vias. In: *Proceedings of IEEE Radio Frequency Integrated Circuits Symposium*, Tampa, FL, pp. 441–444.

[3] Naeemi, A. and Meindl, J. D. 2007. Conductance modeling for grapheme nanoribbon (GNR) interconnects. *IEEE Electron Device Letters* 28(5):428–431.

[4] Li, H., Xu, C., Srivastava, N. and Banerjee, K. 2009. Carbon nanomaterials for next-generation interconnects and passives: Physics, status and prospects. *IEEE Transactions Electron Devices* 56(9):1799–1820.

[5] Kaushik, B. K., Kumar, V. R., Majumder, M. K. and Alam, A. 2016. *Through Silicon Vias: Materials, Models, Design and Performance*. New York: CRC Press.

[6] Kaushik, B. K., Majumder, M. K. and Kumar, V. R. 2014. Carbon nanotube based 3-D Interconnects: A reality or a distant dream. *IEEE Circuits and Systems Magazine* 14(4):16–35.

[7] Chandrakar, S., Gupta, D. and Majumder, M. K. 2021. Role of through silicon via in 3D integration: Impact on delay and power. *Journal of Circuits, Systems and Computers* 30(3):2150051.

[8] Li, H., Xu, C. and Banerjee, K. 2010. Carbon nanomaterials: The ideal interconnect technology for next-generation ICs. *IEEE Design and Test of Computers* 27(4):20–31.

[9] Naeemi, A. and Meindl, J. D. 2007. Design and performance modeling for single-walled carbon nanotubes as local, semi global and global interconnects in gigascale integrated systems. *IEEE Transactions on Electron Devices* 54(1):26–37.

[10] Srivastava, N. Li, H., Kreupl, F. et al. 2009. On the applicability of single-walled carbon nanotubes as VLSI interconnects. *IEEE Transactions on Nanotechnology* 8(4):542–559.

[11] Sarto, M. S. and Tamburrano, A. 2010. Single-conductor transmission-line model of multiwall carbon nanotubes. *IEEE Transactions on Nanotechnology* 9(1):82–92.

[12] Chiariello, A. G., Maffucci, A. and Miano, G. 2013. Circuit models of carbonbased interconnects for nanopackaging. *IEEE Transactions on Components, Packaging and Manufacturing Technology* 3(11):1926–1937.

[13] Kan, E. Li, Z. and Yang, J. 2011. *Graphene Nanoribbons: Geometric Electronic and Magnetic Properties*. In: S. Mikhailov (ed), Rijeka, Croatia: InTechOpen.

[14] Avouris, P. 2010. Graphene: Electronic and photonic properties and devices. *Nano Letters* 10(11):4285–4294.

[15] Jousseaume, V. and Renard, V. T. 2010. Cu based catalysts can make CMOS compatible Si nanowires: Toward reconfigurable interconnects. In: *Proceedings of IEEE International Conference on Interconnect Technology (IITC 2010)*, Burlingame, CA, pp 1–3.

[16] Ni, L., Demami, F., Rogel, R., Salaiin, A. C. and Pichon, L. 2009. Fabrication and electrical characterization of silicon nanowires based resistors. *Materials Science and Engineering: A* 6(1):1–4.

[17] Rakheja, S. and Kumar, V. 2012. Comparison of electrical, optical and plasmonic on-chip interconnects based on delay and energy considerations. In: *Proceedings of IEEE 13th International Symposium on Quality Electronic Design (ISQED 2012)*, Santa Clara, CA, USA, pp. 732–739.

[18] Rashid, Z., Schuller, J., Chandran, A. and Brongersma, M. 2006. Plasmonics: The next chip-scale technology. *Materials Today* 9:20–27.

[19] Zutic, I., Fabian, J. and Sarma, S. D. 2004. Spintronics: Fundamentals and applications. *Reviews of Modern Physics* 76(2):323–410.

[20] Rakheja, S. and Naeemi, A. 2012. Interconnect analysis in spin-torque devices: Performance modeling, optimal repeater insertion and circuit-size limits. In: *Proceedings of IEEE 13th International Symposium on Quality Electronic Design*, Santa Clara, CA, USA, pp. 283–290.

[21] Zhang, G. Warner, J. H., Fouquet, M. et al. 2012. Growth of ultrahigh density single-walled carbon nanotube forests by improved catalyst design. *ACS Nano* 6(4):2893–2903.

[22] Zhao, W. S. Zheng, J., Hu, Y. et al. 2016. High-frequency analysis of cu-carbon nanotube composite through-silicon vias. *IEEE Transactions on Nanotechnology* 15(3):506–511.

[23] Subramaniam, C., Yamada, T., Kobashi, K. et al. 2013. One hundred fold increase in current carrying capacity in a carbon nanotube-copper composite. *Nature Communications* 4 (1): 2202.

[24] Chai, Y. Chan, P. C., Fu, Y. et al. 2008. Electromigration studies of Cu/carbon nanotube composite interconnects using Blech structure. *IEEE Electron Device Letters* 29(9):1001–1003.

[25] Hata, K. Futaba, D. N., Mizuno, K. et al. 2004. Water-assisted highly efficient synthesis of impurity-free singlewalled carbon nanotubes. *Science* 306:1362–1364.

[26] Futaba, D. N. Hata, K., Yamada, T. et al. 2006. Shape-engineerable and highly densely packed single-walled carbon nanotubes and their application as super-capacitor electrodes. *Nature Materials* 5:987–994.

[27] Schlesinger, M. and Paunovic, M. 2010. *Modern Electroplating*. Hoboken: John Wiley & Sons.

[28] Mu, W., Sun, S., Jiang, D., Fu, Y., Edwards, M., Zhang, Y., Jeppson, K. and Liu, J. 2015. Tape-assisted transfer of carbon nanotube bundles for through-silicon-via applications. *Journal of Electronic Materials* 44:2898–907.

[29] Li, H. J. Lu, W. G., Li, J. J. et al. 2005. Multichannel ballistic transport in multiwall carbon nanotube. *Physical Review Letters* 95(8): 086601.

[30] Matsuda, Y., Deng, W. Q. and Goddard, W. A. 2010. Contact resistance for 'end-contacted' metal-graphene and metal-nanotube interfaces from quantum mechanics. *Journal of Physical Chemistry C* 114(41):17845–17850.

[31] Im, S. Srivastava, N., Banerjee, K. and Goodson, K. E. 2005. Scaling analysis of multilevel interconnect temperatures for high-performance ICs. *IEEE Transactions on Electron Devices* 52(12):2710–2719.

[32] Sakurai, T. and Newton, A. R. 1990. Alpha-power law MOSFET model and its applications to CMOS inverter delay and other formulas. *IEEE Journal of Solid-State Circuits* 25(2):584–595.

[33] Xu, C. Li, H., Suaya, R. and Banerjee, K. 2010. Compact AC modeling and performance analysis of through silicon vias in 3-D ICs. *IEEE Transactions Electron Devices* 57(12):3405–3417.

[34] Maffucci, A. 2017. Modeling, fabrication and characterization of large carbon nanotube interconnects with negative temperature coefficient of the resistance. *IEEE Transactions on Components, Packaging and Manufacturing Technology* 7(4):485–493.

[35] Zhao, W. Li, X., Gu, S. et al. 2009. Field-based capacitance modeling for sub-65-nm on chip interconnect. *IEEE Transactions on Electron Devices* 56(9):1862–1872.

[36] Topper, M., Ndip, I., Erxleben, R., Brusberg, L., Nissen, N., Schroder, H., Yamamoto, H., Todt, G. and Reichl, H. 2010. 3-D thin film interposer based on TGV (Through Glass Vias): An alternative to Si-interposer. In: *IEEE Conference on Electronic Components and Technology Conference (ECTC)*, Las Vegas, NV, USA, pp. 66–73.

[37] Sarkar, D., Xu, C., Li, H. and Banerjee, K. 2011. High-frequency behavior of graphene-based interconnects-Part I: Impedance modeling. *IEEE Transactions on Electron Devices* 58(3):843–852.

[38] Zheng, J., -Q, Z., G., Su, Wang, -Y., Li, M., Zhao, W.-S. and Wang, G. 2015. Circuit modeling of Cu/CNT composite through-silicon vias (TSV). In: *IEEE MTT-S International Microwave Workshop Series on Advanced Materials and Processes for RF and THz Applications (IMWS-AMP)*, Suzhou, China, pp. 1–3.

[39] Cheng, D. H. 1993. *Fundamentals of Engineering Electromagnetics* (2nd ed.). Reading, MA: Addison-Wesley.

[40] Sahu, C. C., Chandrakar, S. and Majumder, M. K. 2020. Signal transmission and reflection losses of cylindrical and tapered shaped TSV in 3D integrated circuits. In: *IEEE International Symposium on Smart Electronic Systems (iSES) (Formerly iNiS)*, Chennai, India, pp. 44–47.

[41] Terman, F. 1943. *Radio Engineer's Handbook*. New York: McGraw-Hill.

[42] Grover, F. W. 1946. *Inductance Calculations: Working Formulas and Tables*. New York: Van Nostrand.

[43] Chen, E. and Chou, S. Y. 1997. Characteristics of coplanar transmission lines on multilayer substrate: Modeling and experiments. *IEEE Transactions on Microwave Theory and Techniques* 45(6):939–945.

[44] Stellari, F. and Lacaita, A. L. 2000. New formulas of interconnect capacitances based on results of conformal mapping method. *IEEE Transactions on Electron Devices* 47(1):222–231.

[45] Eo, Y. and Eisenstadt, W. R. 1993. High speed VLSI interconnect modeling based on S-parameter measurements. *IEEE Transactions on Components, Hybrids and Manufacturing Technology* 16(5):555–562.

[46] Kumar, A. and Kaushik, B. K. 2021. Exponential Matrix–Rational Approximation (EM-RA) model for SWCNT bundle and hybrid Cu-CNT interconnects. *IEEE Transactions on Electromagnetic Compatibility* 63(4):1212–1222.

[47] Kumari, B., Kumar, R., Sharma, R. and Sahoo, M. 2021. Design, modeling and analysis of Cu-carbon hybrid interconnects. *IEEE Access* 9:113577–113584.

[48] Sahu, C. C., Anand, S. and Majumder, M. K. 2021. An analysis of the eddy effect in through-silicon vias based on Cu and CNT bundles: The impact on crosstalk and power. *Journal of Computational Electronics* 20:2456–2470.

9 Relative Stability Analysis of the GNR and Cu Interconnect

Sandip Bhattacharya, L. M. I. Leo Joseph,
Sheshikala Martha, Ch. Rajendra Prasad,
Syed Musthak Ahmed, Subhajit Das,
Debaprasad Das, and P. Anuradha

9.1 INTRODUCTION

Future challenges with on-chip interconnect dependability and durability are being anticipated by the interconnect technology's rapid advancement. In sub-nanometer designs, the interconnect stability declines with decreasing connector dimension. Traditional copper-based interconnects are less thermally viable and suffer from electromigration issues, surface scattering, and grain boundary scattering at the nanoscale size. A suggestion has been made for the graphene nanoribbon (GNR) as a potential innovative material for future technological nodes to solve such issues [1–5]. Due to its high current density, large electron mobility, large electron mean free path (MFP), and good thermal conductivity, graphene nanoribbon (GNR) can take the role of copper, for nano-interconnect [3,5]. In general, multilayer GNR structures are preferable over single-layer GNR structures for simulating nano-interconnects because single-layer GNR exhibits high electrical resistance [3]. According to [3,4,6,7], three resistances are crucial in monolayer GNR: quantum resistance (R_Q), scattering resistance (R_S), and contact resistance (R_C). R_S values rely on mean free path $(\lambda_{effective})$, Fermi energy (E_F), interconnect length (l), and connector width (w). A multilayer GNR stability model was initially put forth by Nasiri et al. using a Nyquist plot [2]. Using multilayer GNR interconnects, a crosstalk stability study is presented in [8]. The interconnect delay and Nyquist stability analysis of multilayer GNR and MWCNT interconnects has been presented by Koushik et al. [9]. In this study, we give a comparative stability analysis of the TC-GNR, SC-GNR, and Cu interconnect. Here, the analysis is performed using Nyquist and Bode plot to examine the relative stability for TC-GNR, SC-GNR, and Cu interconnects in response to prior works [2,8–13]. According to our investigation, the SC-GNR interconnect exhibits greater relative stability when compared to the Cu and TC-GNR interconnects. Furthermore, it is discovered that the relative stability of Cu is insignificant compared with TC-GNR and SC-GNR. It has been demonstrated to be superior to SC-GNR and TC-GNR connection across various technological nodes when taking into account the other advantages of Cu interconnect.

DOI: 10.1201/9781003331650-9

9.2   FORMULATION OF TRANSFER FUNCTIONS

The interconnect system's transfer function was created using a comparable circuit model for the GNR (TC and SC) and Cu interconnects, as illustrated in Figures 9.1 and 9.2, in order to explore the stability of all three interconnect systems. The series connection of RLC value makes up the RLC equivalent circuit model. The unit step response $u(t)$ of the interconnect systems is shown in Figure 9.3. The RLC circuit's transfer function is provided by

$$H(S) = \frac{v_{out}(s)}{v_{in}(s)} = \frac{1}{LC.s^2 + RC.s + 1} \tag{9.1}$$

Complex frequency is defined as $s = \sigma + j\omega$, where R_C is a series of imperfect contact resistance, R_O is denoted by ohmic resistance, R_Q is denoted by quantum resistance, L_K is a series of kinetic inductance, L_M is represented as magnetic inductance, C_E is a series of electrostatic capacitance, and C_Q is denoted by the quantum capacitance.

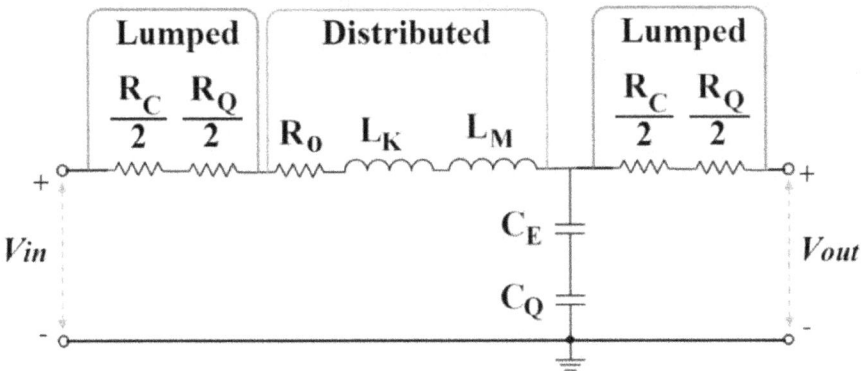

FIGURE 9.1   MLGNR interconnects' equivalent circuit model.

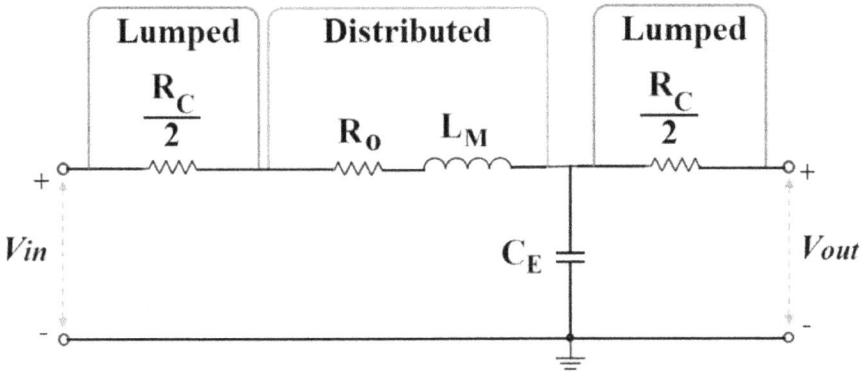

FIGURE 9.2   Cu interconnects' equivalent circuit model.

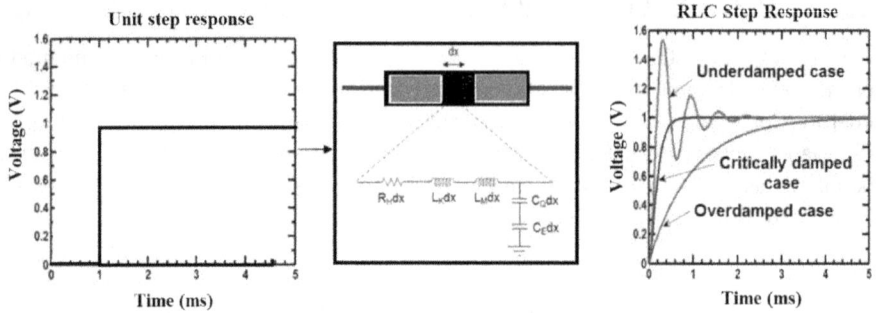

FIGURE 9.3   Stability response analysis using unit step function [13].

For varying interconnect lengths (*l*) and different interconnect widths (*w*), the RLC parameters for three different interconnects are obtained from [7]. The transfer function (TF) of all interconnect systems was determined by substituting the RLC values into equation (9.1). After determining the transfer function, a unit step input is used to perform the Bode analysis approach to verify the relative stability. To verify the relative stability study, the Nyquist analysis is carried out for both the Cu interconnect and GNR (TC and SC).

9.3   ANALYSIS OF BODE STABILITY

A model depicted in Figure 9.3 was taken into consideration to examine the Bode stability analysis for an LTI system. Here, a unit step input is provided as shown in Figure 9.3, and three different types of responses are observed: At the system's output, (1) overdamped, (2) underdamped, and (3) critically damped are presented.

   The gain and phase margin (GM and PM) of a system are two ways to gauge Bode stability. The stability of the system increases as the gain and phase margin are raised. While the *X*-axis in a Bode plot shows frequency, the *Y*-axis in a Bode plot represents phase (in deg) and amplitude (in dB). The interconnect system's gain and phase margin were determined using the MATLAB tool. The amount of gain that may be estimated from the phase crossover frequency to the 0 dB line is known as the gain margin (GM). The gain curve's cross point at 180 and the phase at 0 dB is known as the phase margin (PM) [14]. In this work, by extending the interconnect length (*l*) from 1 to 100 μm and changing connection widths (*w*), we have investigated the relative stability of the three distinct types of interconnect systems. The network delay increases along with the link length. The unit step response of three different types of interconnect systems tends to dampen more quickly as the switching delay grows, and the system becomes more stable [2,14].

9.4   ANALYSIS OF NYQUIST STABILITY

For additional validation, the same model is subjected to the Nyquist stability analysis method. We can further confirm whether or not our earlier model produces the same response using this stability study. A parametric plot of a transfer function called a Nyquist graph is employed in the control system. Nyquist graphs are most frequently used to assess the stability of open and closed-loop feedback systems. The real part of the transfer function (TF) is shown by a plot on the X-axis in Cartesian dimensions. On the Y-axis, the imaginary part is displayed. A graphic is produced for every frequency when the frequency is scanned as a parameter. In contrast, when plotting a transfer function in polar coordinates, the gain of the function is represented by the radial coordinate, and the phase by the angular coordinate [14,15]. Applying the Nyquist stability criterion to the Nyquist plot of the open-loop system allows for the evaluation of the stability of a closed-loop negative feedback system (i.e., the same system without its feedback loop). Even systems with delays and other non-rational transfer functions, which may seem challenging to evaluate using conventional methods, can be easily applied using this method. A measure of stability is the number of times the location has been encircled (1,0). By observing the crossing of the real axis, it is possible to establish the range of gains across which the system will be stable. The Nyquist plot can reveal some details about the transfer function's shape. By measuring the angle at which the curve approaches the origin, for instance, the plot can reveal information on the difference between the transfer function's [14,15] number of poles and zeros.

9.5   RESULTS

In this section, we analyzed the relative stability of three different interconnect materials using Bode and Nyquist stability analysis. In order to create alternative Bode plots and Nyquist plots, we changed the length of the three different interconnects (Cu, TC-GNR, and SC-GNR) for 16 nm technology nodes from 1 to 100 μm. We have investigated the relative stability of the system utilizing various connecting materials, as shown in Figures 9.4–9.6, based on the Bode plot. The Bode plot mainly consists of magnitude (dB) and phase (degree) response w.r.t. frequency. Three distinguished lengths (10, 50, 100 μm) and 16 nm interconnect width is considered for Bode plot analysis.

Different interconnect systems' gain margins and phase margins are displayed in Table 9.1. When comparing gain margin values between different interconnect lengths from 1 to 100 μm, it is found that SC-GNR interconnects exhibit a greater gain margin value than TC-GNR and Cu interconnects. The maximum gain margin is observed at 1 μm (310 dB) interconnect for SC-GNR interconnect. The SC-GNR exhibits constant stability (green color) and a greater gain margin from Figures 9.4–9.6. According to the aforementioned analysis, SC-GNR can be used as an interconnecting material in nano-electronics applications where more stability is needed. Nyquist plot or graph was also used to investigate the relative stability of the TC-GNR, SC-GNR, and copper interconnects (Figures 9.7–9.9). Because there isn't a pole on the right-hand side of the S plane and because of how many times the point is encircled, the SC-GNR and TC-GNR connections mostly exhibit stability (1, 0). The system will become unstable beyond real axis −1. Our analysis reveals that, in

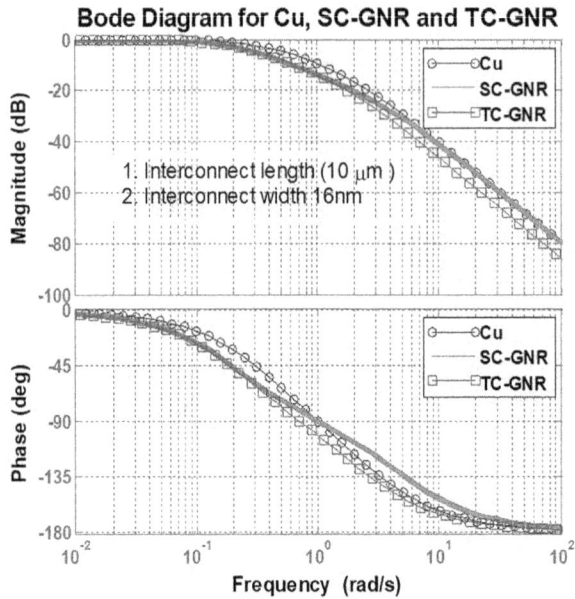

FIGURE 9.4 Phase vs frequency and magnitude vs frequency plot at 10 μm interconnect length.

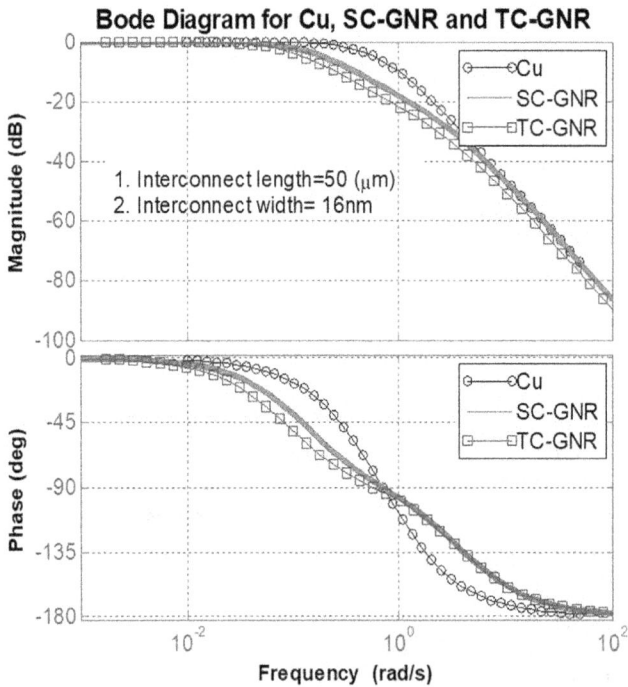

FIGURE 9.5 Phase vs frequency and magnitude vs frequency plot at 50 μm interconnect length.

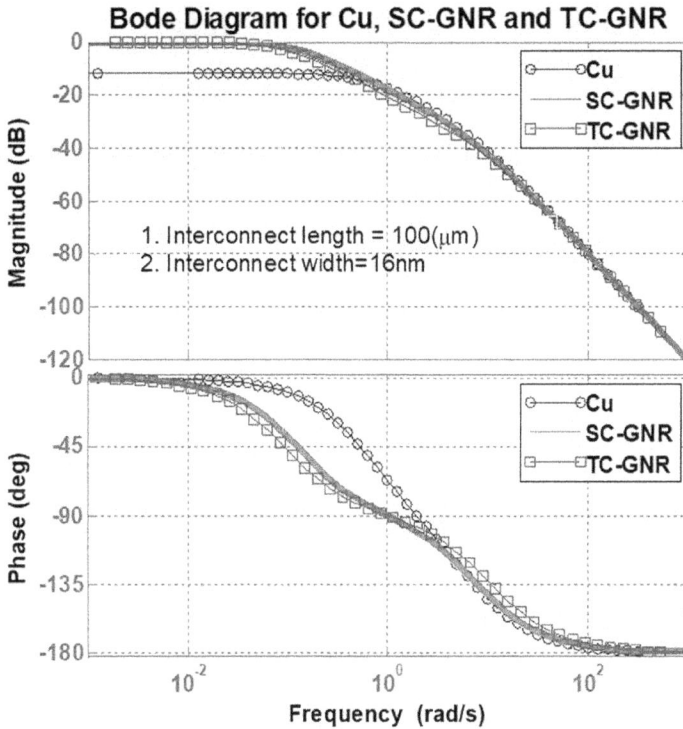

FIGURE 9.6   Phase vs frequency and magnitude vs frequency plot at 100 µm interconnect length.

TABLE 9.1
Interconnect System with GM (in dB) and PM (in deg)

Interconnect Length (µm) →	1	5	10	50	100	1	5	10	50	100
Types of Interconnects ↓			GM					PM		
Cu	285	282	281	279	278	90	90	90	90	90
TC-GNR	299	298	293	291	290	90	90	90	90	90
SC-GNR	310	307	306	305	302	90	90	90	90	90

contrast to the Cu interconnect, where the majority of the encircle passes beyond the −1 at the real axis (see Figures 9.9), all encircles for SC-GNR and TC-GNR pass through the origin with varied interconnect lengths as shown in Figures 9.7 and 9.8. The fact that all lines are passed beyond the −1 line in the real axis indicates that Cu interconnect is an unstable material for use in next-generation nano-electronic circuit design. When the length is equivalent to 1, 5, and 10 µm, SC-GNR and TC-GNR

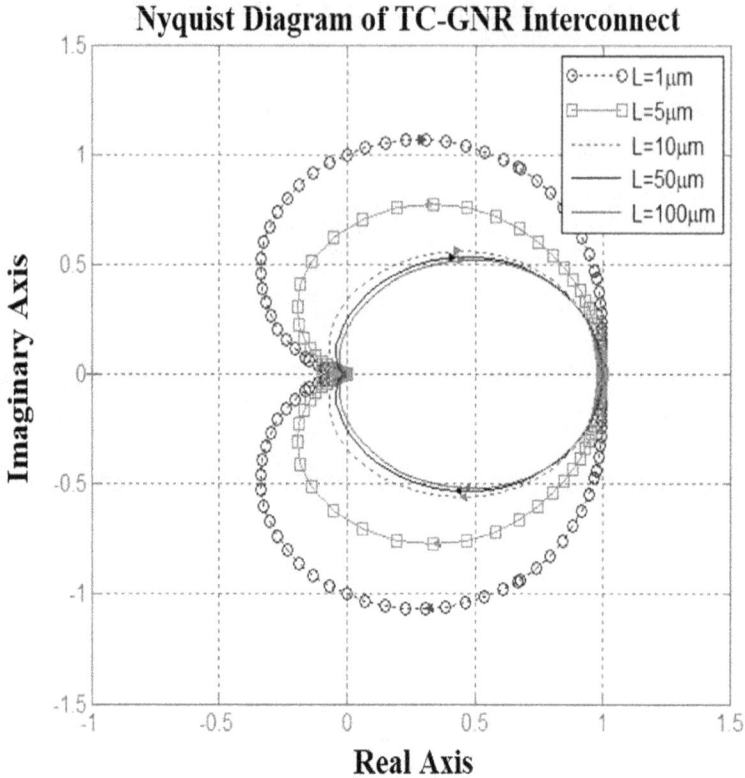

FIGURE 9.7 Real axis (*X*-direction) and imaginary axis (*Y*-direction) plot for SC-GNR interconnect (*l* = 1–100 μm).

interconnect exhibit poor stability in both circumstances. The encirclements of the contour present between −1 and 0 on the real axis indicate some stability if we wish to expand the interconnect length of GNR from 50 to 100 μm.

9.6 CONCLUSIONS

In this study, we compared the stability of the TC-GNR, SC-GNR, and Cu interconnects for nodes using the 16 nm ITRS technology with varying interconnect lengths (1–100 μm). We have demonstrated from the Bode graphs that relative stability reduces with increasing connection length. This is because the lengthening of the system interconnects will result in longer interconnect delays. As a result, the system's step reaction tends to become slower in nature (overdamped), which makes the system more unstable.

Nyquist Diagram of SC-GNR Interconnect

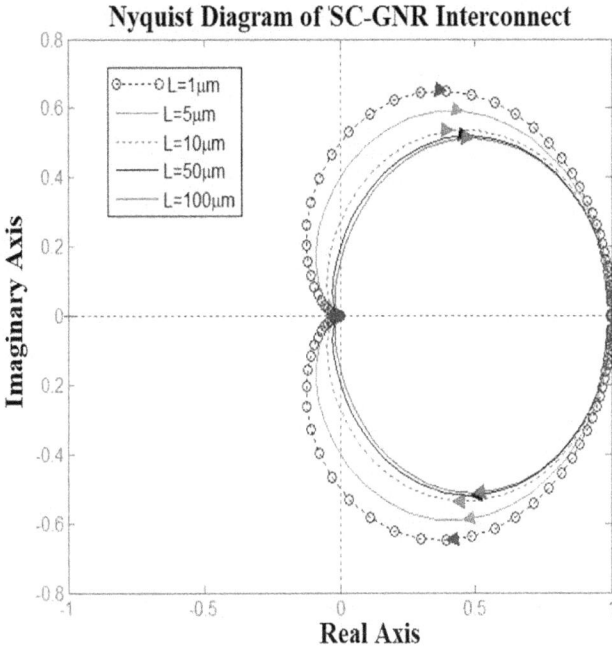

FIGURE 9.8 Real axis (*X*-direction) and imaginary axis (*Y*-direction) plot for TC-GNR interconnect ($l = 1$–100 μm).

Nyquist Diagram of Cu Interconnect

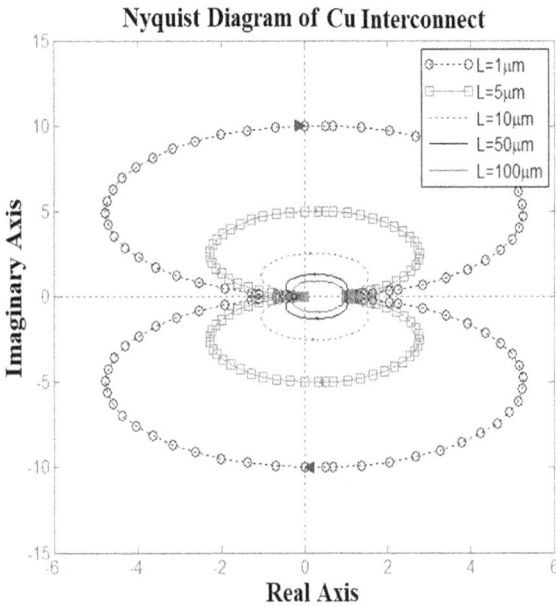

FIGURE 9.9 Real axis (*X*-direction) and imaginary axis (*Y*-direction) plot for SC-GNR interconnect ($l = 1$–100 μm).

REFERENCES

[1] C. Xu, H. Li, and K. Banerjee. Modeling, analysis, and design of graphene nano-ribbon interconnect. *IEEE Transactions on Electron Devices*, 56(8):1567–1578, 2009.

[2] S.H. Nasiri, M.K. Moravvej-Farshi, and R. Faez. Stability analysis in graphene nanoribbon interconnects. *IEEE Electron Device Letters*, 31(12):1458–1460, 2010.

[3] A. Naeemi, and J.D. Meindl. Compact physics-based circuit models for graphene nanoribbon interconnects. *IEEE Transactions on Electron Devices*, 56(9):1822–1833, 2009.

[4] A. Naeemi, and J.D. Meindl. Conductance modeling for graphene nanoribbon (GNR) interconnects. *IEEE Electron Device Letters*, 28(5):428–431, 2007.

[5] K.I. Bolotin, K.J. Sikes, Z. Jiang, M. Klima, G. Fudenberg, J. Hone, P. Kim, and H.L. Stormer. Ultrahigh electron mobility in suspended graphene. *Solid State Communications*, 146(910):351–355, 2008.

[6] A.K. Nishad, and R. Sharma. Analytical time-domain models for performance optimization of multilayer gnr interconnects. *IEEE Journal of in Selected Topics in Quantum Electronics*, 20(1):17–24, 2014.

[7] S. Bhattacharya, D. Das, and H. Rahaman. Reduced thickness interconnect model using gnr to avoid crosstalk effects. *Journal of Computational Electronics*, 15(2):367–380, 2016.

[8] L. Akbari, and R. Faez. Crosstalk stability analysis in multilayer graphene nanoribbon interconnects. *Circuits, Systems, and Signal Processing*, 32:2653–2666, 2013.

[9] V.R. Kumar, M.K. Majumder, A. Alam, N.R. Kukkam, and B.K. Kaushik. Stability and delay analysis of multi-layered GNR and multi-walled cnt interconnects. *Journal of Computational Electronics*, 14:611–618, 2015.

[10] S. Bhattacharya, S. Das, and D. Das, Analysis of stability in carbon nanotube and graphene nanoribbon interconnects. *International Journal of Soft Computing and Engineering* 2(6):325–329, 2013.

[11] S. Das, S. Bhattacharya, D. Das, and H. Rahaman, RF performance analysis of graphene nanoribbon interconnect. In: *Proceedings of the 2014 IEEE Students' Technology Symposium (TechSym)*, IIT Kharagpur, 2014, pp. 105–110.

[12] S. Das, S. Bhattacharya, D. Das, and H. Rahaman, Thermal stability analysis of graphene nano-ribbon interconnect and applicability for terahertz frequency. *National Academy Science Letters* 43(3):253–257, 2020.

[13] S. Bhattacharya, D. Das, and H. Rahaman, Stability analysis in top-contact and side-contact graphene nanoribbon interconnects. *IETE Journal of Research* 63(4):588–596, 2017.

[14] K. Ogata. *Modern Control Engineering*, 5th ed., Pearson Education, 2015.

[15] R.C. Dorf, and R.H. Bishop, *Modern Control System*, 11th ed., Englewood Cliffs, NJ: Prentice-Halls, 2008.

10 Transmission Line-Based Modeling of CNT and GNR Interconnects Using Numerical Methods

Shashank Rebelli

10.1 INTRODUCTION

10.1.1 BACKGROUND

The density and complexity of very-large-scale integrated (VLSI) circuits has increased exponentially over the last two decades resulting in high-performance electronic systems for a wide range of applications such as reconfigurable computing, mobile and satellite communication, multimedia, micro-electromechanical systems (MEMS), and robotics. The count of active devices has reached hundreds of millions, while connecting wires among the devices tend to grow linearly with transistor counts [1].

An integrated circuit (IC) comprises of several components and functional blocks, such as transistors, gates, and sub-circuits, which are interconnected using aluminum (Al)/copper (Cu) metals or graphene-based materials. Interconnects are capable of transmitting data from one block to the other, in the form of current or voltage. Ideally, the signal transmission/reception between the two interconnected blocks should be instantaneous with no delay. However, this cannot be achieved in practical situations, due to the fact that there always exists a signal propagation time during the transmission of data from one block to the other. If the signals vary rapidly (high-frequency applications) compared to the propagation time, several effects may be observed such as Delay, Overshoot, and Crosstalk [2]. Currently in the deep submicron (DSM) regime, performance of electronic systems depends on these effects introduced by interconnections. Hence, it is very important to have accurate and efficient estimation models of the interconnection effects at the design phase itself to avoid pitfalls and to reduce the time to Market of VLSI chips.

In the state of the art, there are many viable models developed (with variable degrees of accuracy), starting from a simple capacitor model to a frequency-dependent transmission line (TL) model. These models are more or less a simplified analysis of the fundamental physical event, i.e., the propagation of an electromagnetic (EM) wave in the complex metal-dielectric structure formed by the interconnection network of an electronic circuit. Here, the complication is that a comprehensive EM

DOI: 10.1201/9781003331650-10

analysis of an IC is beyond the present-day computation capabilities. Hence, the EM phenomena are replaced with electrical models. Even complex electrical models are replaced with simpler electrical models when accuracy is not critical, because the simulation time of complete system would otherwise be extremely long [3].

The length of an interconnect determines its classification as either local, intermediate, or global [4]. Thin wires called "local interconnects" are required to connect gates as well as transistors within the exact same block on a circuit. This variety of interconnects generally occupied the lower few metal layers in a multi-layered interconnect system as depicted in Figure 10.1. Clock and data signal distribution inside a functional block or between adjacent blocks, where the standard length between blocks can range from 0.5 to 2.5 mm, is accomplished by the use of intermediate (semi-global) interconnects. These types of interconnects typically exist a few layers above local interconnects. Global interconnects are used to connect a large number of intellectual property (IP) blocks, such as filters, memory, processing elements, and

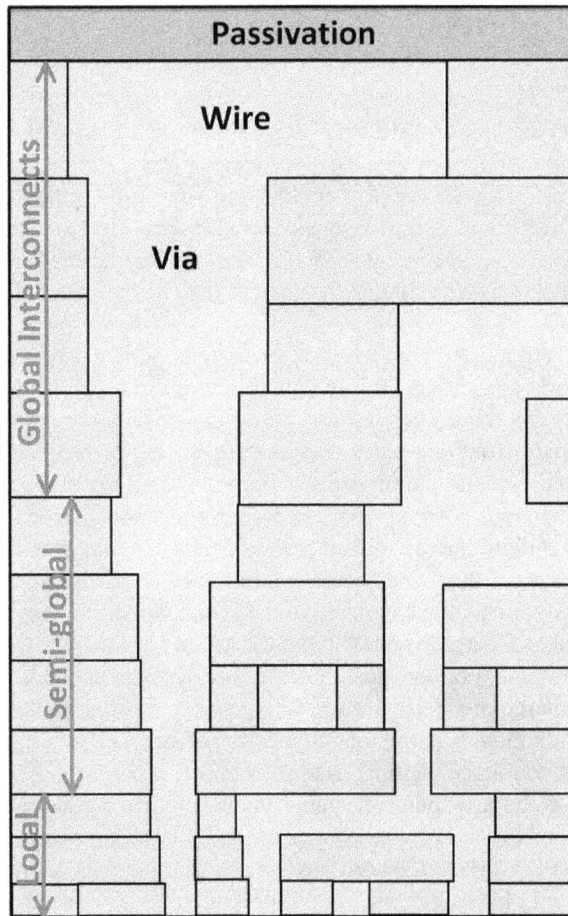

FIGURE 10.1 Sketch of an IC cross section showing its layer stack [1].

interfaces. As these IP blocks need to interact with each other over great distances, they need long wires that span most of the length of the entire chip size. Wide and long metal layers serve the global interconnects, which fill the top few layers of a multi-layer system. These global interconnects typically span over 2.5 mm and can extend to as much as half the chip's perimeter in extreme cases. In most cases, the upper metal layers are used to route power, ground, and clock signals. The dimensions of local interconnects perfectly scale with technology scaling; however, the dimensions of intermediate and global interconnects do not scale proportionally with technology scaling.

With technology scaling the gate and wire delay of the local interconnects decreases, whereas the delay of the semi-global/global interconnects increases. Advanced scaling techniques such as the use of low K dielectric materials improve the delays of the local interconnect. However, with the scaling of every successive technology node below 0.25 μm, the delays of semi-global/global interconnect are more detrimental than gate delays [1]. In addition, as compared to the local interconnect, the length of global interconnect is not scaled with technology, leading to an increase in delays as these wires need to run across the entire chip. Thus, as compared to the gate delay and local interconnect delay the global interconnect delay becomes a limiting factor in determining the overall circuit performance in the present-day VLSI chips. Global wiring among the functional blocks provides the distribution of clock/signal and delivers ground/power to all functions on an IC. Figure 10.2 illustrates the local and global interconnect delay in future generations. To reduce the delay in global wiring, the repeaters can be incorporated by compromising the power consumption and chip area.

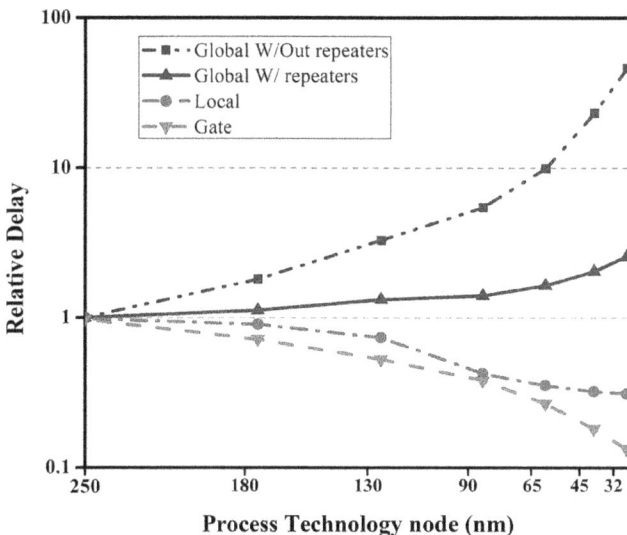

FIGURE 10.2 On-chip interconnect delay at various stages of technology advancement adopted from [1].

In DSM VLSI regime, the speed of any electrical signal depends on two factors, namely the transistor gate delay and the propagation delay of interconnects. The major challenges in VLSI circuits are global interconnect delays and crosstalk noise between multiple interconnects. These signal integrity (SI) issues in interconnects decide the overall performance of VLSI circuits. For the iterative layout design of densely populated ICs, accurate analytical models are needed to efficiently predict signal degradation due to propagation delay, crosstalk noise, and signal overshoot in the early design cycles [5–7]. The existing tools like computer-aided design (CAD) for SI analysis are more time-consuming and inefficient. Hence, interconnect simulations suffer from a number of SI problems that require advanced CAD tools for analysis. Computationally high-speed and accurate interconnect models are needed at the initial stages of an IC design for design optimization and post-layout verification, respectively. During the physical design, interconnect area, propagation delay, overshoot, power, and crosstalk noise estimations are the main performance metrics. This chapter addresses the estimation of propagation delay and crosstalk noise in the mutually coupled on-chip interconnects.

10.1.2 MATHEMATICAL MODELS FOR ESTIMATION OF DELAY AND PEAK CROSSTALK NOISE IN THE INTERCONNECTS: A TECHNICAL REVIEW

In the early phase, the gate capacitance of transistors dominated the interconnect parasitic capacitance, which makes use of the assumption of modeling of interconnects as short circuits. Later on, with technology scaling, the interconnect parasitic capacitance dominates the gate capacitance and interconnect was modeled as a lumped capacitance [8,9]. With the further technology downscaling, inclusion of resistance effect in on-chip interconnect became mandatory for global interconnects, which increases the accuracy. This results in introduction of lumped resistance-capacitance (RC) models for the performance analysis of on-chip interconnects [10,11]. However, the lumped RC models are treated as the distributed RC model [5] for better accuracy. Currently, because of the high switching frequencies and the adoption of low-resistive interconnect materials, the parasitic inductance plays an important role in the performance of on-chip interconnects. To estimate the performance of the interconnects accurately, they must be considered as transmission lines or as distributed resistance inductance capacitance (RLC) interconnect lines [12].

Initially, crosstalk noise estimation models considered only capacitive coupling [13,14]. However, inductive-crosstalk effects should be included at current high-frequency operations for the inclusive analysis of coupling noise. At high frequencies, the transient crosstalk, i.e., the undesired effect of a signal transmitted on one line over another, is produced due to closely packed interconnects [15–17]. The propagation delay of the signal is strongly influenced by the crosstalk noise, which results in functional failure or circuit malfunction. The crosstalk between the coupled lines is considered to be dynamic and functional, depending on the input switching transitions in the coupled interconnects. Dynamic crosstalk occurs when the adjacent lines are simultaneously switching either in-phase or out-of-phase, whereas the functional crosstalk appears as a voltage spike when the victim line is quiescent while

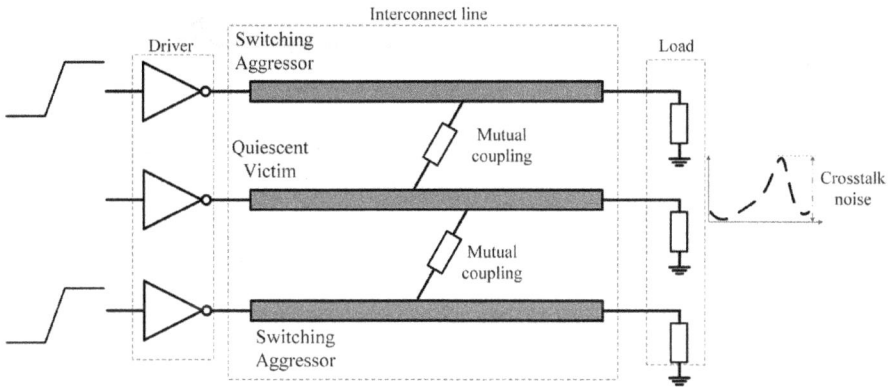

FIGURE 10.3 Crosstalk in closely placed interconnects.

switching an aggressor line as shown in Figure 10.3. A change in propagation delay and logic value can be observed under dynamic and functional crosstalks, respectively. In addition, the effect of crosstalk noise includes ringing and signal overshoot/undershoot. Therefore, there is a need for an accurate estimation of performance parameters for designing high-performance on-chip interconnects incorporating the effect of crosstalk noise.

A noise model [18] was proposed for the analysis of noise effects in two coupled RLC lines, but it is limited to loosely coupled interconnect lines where coupling capacitance and mutual inductance are negligible as compared to ground capacitance and self-inductance, respectively. Similarly, another analytical model [19] was proposed for coupled on-chip RLC line, in which two lines were isolated. Further, each isolated line is approximated as a one segment RLC π-circuit. The major limitation of this method is that it can be applied only to isolated lines with separated drivers. Agarwal et al. [20] proposed an analytical scheme to model crosstalk noise in the coupled RLC interconnects by considering the linear characteristics of CMOS driver, i.e., linear resistive driver. This model is further extended to a nonlinear CMOS driver considering α-power law model to analyze dynamic crosstalk effects [21] and functional crosstalk effects [22] by Kaushik et al. The models that are purely limited to two coupled interconnect lines are based on even-odd modes reported in [20–22]. Furthermore, only the ideal or lossless lines are considered for transient analysis.

10.1.2.1 Analytical Models

In 1948, Elmore [10] developed an analytical model based on the first moment for estimating the delay of amplifier circuits, later this model was used for fast extraction of delay in simple RC interconnects. For ICs composed of millions of gates, it is often impractical to use highly accurate and computationally efficient models to estimate delay at each and every node in the circuit. Hence, the Elmore delay model is used as a quick estimator of relative delay calculation of different paths in the circuit. Figure 10.4 shows an interconnect tree network composed of resistance and capacitance elements. The delay of any path of tree network using this model is written as equation (10.1).

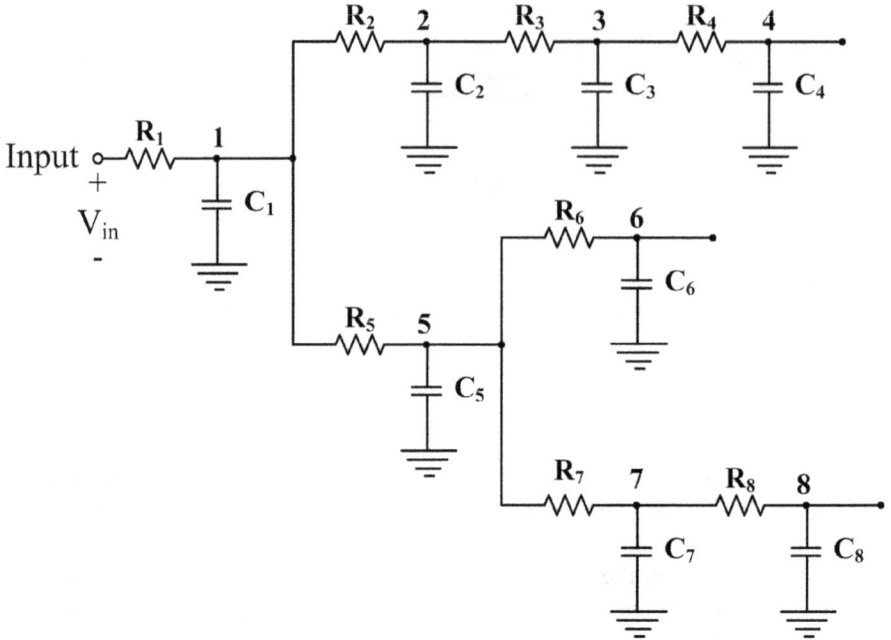

FIGURE 10.4   RC interconnect tree network.

$$\tau_{DK} = \sum_{j=1}^{K} C_j R_{jj} \Rightarrow \tau_{Dj} = C_1 R_1 + C_2\left(R_1 + R_2\right) + \cdots + C_j\left(R_1 + R_2 + \cdots + R_j\right) \quad (10.1)$$

For noise analysis between two coupled lines, Sicard and Rubio [23] proposed a simplified model, which evaluates the effects of parasitic capacitive coupling. But this is applicable only for the interconnects when they are considered as simple capacitor models. Also, Vittal and Malgorzata [24] have considered the lumped capacitor model for appropriate channels and derived bounds for noise expressions using the lumped model where the line resistance was ignored. Later, this work is further extended in [14] to incorporate lumped π-model for RC interconnects. Besides, the extension of these generalized expressions to the distributed models is less complicated.

In order to achieve better accuracy the lumped RC model must be treated as the distributed RC model. Sakurai [11] provided analytical delay calculations for distributed RC interconnects. This model uses the Heaviside expression for calculation of time domain response. Using these expressions, the optimized width is calculated to reduce the bus RC delay. This optimized width is about half of the pitch provided the pitch is less than four times the height. For various source and load capacitance values, 90% and 50% delay values were presented in equations (10.2) and (10.3), respectively. The heuristic delay equation in this model is identical to the Elmore delay equation, which has similar constraints of the Emore delay model [10].

$$\frac{t_{09}}{RC} = 1.02 + 2.3\left(R_T C_T + R_T + C_T\right) \quad (10.2)$$

$$\frac{t_{05}}{RC} = 0.377 + 0.693(R_T C_T + R_T + C_T) \tag{10.3}$$

Moreover, Sakurai [11] described the step response of the distributed RC network as a power series, and derived an expression to analyze the coupled noise in the victim line. This model extends the equation for the voltage at the end of two coupled lines to first order since the series is too complex to calculate analytically.

Currently, because of the high operating switching frequencies and the adoption of low-resistive interconnect materials, the parasitic inductance plays an important role in the performance of on-chip interconnects. To estimate the performance of the interconnects accurately, they must be considered as TLs or as distributed RLC lines [12].

Based on the moments of the first and second orders, Kahng and Sudhakar [25] proposed an analytical delay model for distributed RLC lines under step input to include the effects of inductance. The estimated delay using this model is within 15% of SPICE delays. They also extended their model to estimate the delay values for various combinations of source and load parameters. Yu et al. [26] developed a novel analytical approach, which is a second-order RLC interconnect model for estimating delay, crosstalk noise, and overshoot accurately. Depending on current return path identification, this model can be used to decouple a set of coupled interconnects. Ismail et al. [27] developed an equivalent Elmore model for estimation of 50% delay in an RLC interconnect tree. This closed-form delay model includes all damping conditions (both monotonic and non-monotonic nature) of an RLC interconnect, which mainly differs from the Elmore delay model. This model provides closed-form solutions for the settling time, rise/fall time, 50% delay, and overshoot of signals in the distributed RLC interconnect tree. Out of these expressions, the delay formula of an RLC interconnect tree has similar accuracy characteristics with respect to the Elmore delay model. Davis and James [28] proposed a new compact model for the accurate analysis of transient response, overshoot, and delay in the single high-speed distributed RLC interconnect, and the same is extended to coupled interconnect lines [29] for accurate estimation of peak crosstalk noise and transient response. Another efficient coupled crosstalk noise estimation method based on the model-order reduction approach is developed by Martin and Sachin [30]. This method computes the noise according to the time constant of the aggressor signal, the conductances and sink capacitances of the victim and the aggressor nets, respectively, and the coupling capacitance between those two nets.

Moreover, most of the researchers described the distributed RLC interconnect line as a transmission line based on the ABCD matrix approach [2, 31–35]. Banerjee and Amit [31] introduced an efficient analysis of inductance effects for global interconnects and the time domain response of a DIL system. In this model, interconnect is driven by a series resistance and output parasitic capacitance of a repeater and the same is terminated by a load capacitance. Using the ABCD matrix approach and Laplace domain techniques, they have presented the accurate expressions for the transfer function of these global interconnect lines and their delay calculations.

Li et al. [32] derived a new recursive model considering the ABCD matrix of transmission lines for accurate estimation of propagation delay and time domain response of interconnect trees. This method provides the exact transfer function of a distributed RLC interconnect tree using second-order approximation and moment matching for propagation delay calculations and fast simulation time, respectively. The accuracy of the method is validated by comparing it with HSPICE simulations.

Palit et al. [33] presented the method of ABCD modeling, which is needed to model the coupling noise on the victim line due to single or multiple aggressors. The same author in another contribution [34] developed a distributed RLGC and decoupled victim line model by considering all possible sources or coupling noise (mutual inductance and mutual capacitance between the two adjacent victim and aggressor lines).

Zhou et al. [35] presented an RLC model to improve the accuracy of interconnect delay predictions in ICs. Initially, using the ABCD matrix the first two moments of the circuit are derived. And then the total delay is estimated using the rise time delay and transport delay. Another analytical model based on a Fourier series representation of the periodic input signal is proposed by Chen and Friedman [2], for the estimation of delay in RLC interconnect trees and the analysis of crosstalk noise in multiple lines. In this model, a transfer function of the interconnect line is derived based on the ABCD parameters.

However, with higher clock frequencies, the on-chip interconnect lines behave as lossy TLs. Therefore, some researchers estimated coupling noise and delay of on-chip interconnects by adopting the transmission line model. Agarwal et al. [20] developed an analytical framework based on transmission line theory to model coupling noise in coupled RLC lines. This model is applied to coupled lines under the terminal conditions introduced by CMOS drivers and receivers. But, the nonlinear effects of CMOS driver are replaced with their equivalent linear resistor and a capacitive load is considered at the receiver. This model is further extended by Kaushik and Sarkar [22,36] to include the nonlinear characteristics of the CMOS driver. These authors have adopted α-power law model [37] to represent the CMOS driver characteristics. Then, Kaushik et al. [21] proposed the transmission line model with α-power law model for DIL system to analyze the dynamic in-phase and out-of-phase delay in coupled lines.

These models that are purely restricted to two coupled lines based on even-odd modes are reported in [20–22]. The modeling of CMOS driver-based distributed RLC lines suffers from problem of time/frequency domain conversion. This issue occurred due to the CMOS driver being modeled in the time domain, whereas to solve the TL model in the frequency domain, the partial differential equations are used. Therefore, many researchers substitute the nonlinear CMOS driver with a linear resistive driver compromising the overall accuracy of the model. In the recent past, the TL equations in time domain were solved using the FDTD model [24] to avoid the conversion problem.

10.1.2.2 Numerical Models: FDTD Method

The FDTD model is a suitable numerical approach for computational EM modeling. Initially, this method was developed by K. S. Yee [38] to solve Maxwell's equations in the time domain by discretizing the time-dependent PDEs in time and space considering central difference schemes. Later in 1994, C. R. Paul [39] analyzed multiconductor TLs by incorporating lumped boundary conditions into the FDTD method. Then, this work was extended to analyze the lossy TLs by Roden et al. [40]. Similarly, Li et al. [41] presented accurate numerical FDTD method for the analysis of the transient response of a single TL driven by CMOS inverter driver, which exhibits better accuracy with respect to SPICE simulation results. Therefore, the FDTD method with better accuracy has attracted many researchers to analyze the SI issues in high-speed interconnects.

Li et al. [24] presented FDTD-based model for estimating the transient response of CMOS gate-driven coupled RLGC interconnects at 180 nm technology node. This work presents the FDTD model with a second-order accuracy to solve telegrapher's equations, for analysis of coupled interconnects and the nonlinear behavior of the CMOS gates is modeled using α-power law model. Sharma et al. [42] adopted the FDTD method for both dynamic and functional crosstalk analysis in lossy RLC interconnects. The authors validated the computed results with respect to HSPICE simulations. But they considered the linear resistive driver to drive the interconnects. In [43], they have studied the effect of coupling parasitics on propagation delay and coupling noise in the coupled RLC lines.

Recently, Kumar et al. [44] developed an FDTD model for accurate analysis of dynamic crosstalk in CMOS gate-driven coupled on-chip Cu interconnects at 130 nm technology node. And the CMOS gate was represented using α-power law model. However, α-power law model becomes imprecise as the technology scales down below 180 nm, as the finite drain conductance parameter (λ_d) is ignored in the α-power law model. Therefore, in [45] they proposed the FDTD scheme for the analysis of on-chip Cu interconnects at 32 nm technology node, which uses the nth-power law model [46] to represent the nonlinear driver. The nth-power law model is more accurate as it includes the λ_d and velocity saturation effect. Later on, Kumar et al. analyzed the inclusive crosstalk noise of coupled MWCNT interconnects driven by linear driver (resistive) [47] and the same is extended to nonlinear driver (CMOS) [48] considering modified α-power law model. They developed a modified α-power law model such that the effect of λ_d is included. Further, they applied the same to MLGNR interconnects [49].

Similarly, Agarwal et al. [50] proposed an accurate model based on the FDTD model to analyze the crosstalk noise effects in current-mode signaling (CMS) and voltage-mode signaling (VMS) coupled on-chip interconnects at 32 nm technology node. They represented nonlinear characteristics of CMOS gate with nth-power law model. Later, the same is extended to MWCNT [51] and MLGNR [52] interconnects. Recently, Amit et al. [53] adopted the FDTD method for transient analysis of crosstalk effects and temperature-dependent equivalent single conductor (ESC) modeling for mixed-carbon nanotube bundle (MCB) interconnects.

Because of its exceptional accuracy, the FDTD method has been widely accepted by researchers as an important numerical approach for dealing with EM problems and PDEs. Although the FDTD approach is accurate, it causes an error due to numerical dispersion [49, 50] during the propagation along the discretization. Due to this, there is a great need for a cutting-edge, new approach that is effective at handling numerical dispersion characteristics.

10.1.2.3 Numerical Model: MRTD Method

The MRTD scheme is a numerical method based on wavelets proposed by Katehi and Krumpholz [54], which provides an efficient model for computation of EM field. The numerical dispersion of the MRTD algorithm has shown great efficiency to approximate the most accurate solution with negligible error as compared to the FDTD model. They have considered wavelet functions and cubic spline Battle-Lemarie scaling to derive the MRTD algorithm. The dispersion curves of the MRTD model based on Battle-Lemarie scaling function compared with ideal case and FDTD scheme are illustrated in Figure 10.5. Tentzeris et al. [55] performed dispersion and stability analysis of MRTD algorithm based on Battle-Lemarie scaling function for zero-resolution wavelets and different stencil sizes.

In [57] and [58], the MRTD algorithm was derived considering Haar scaling and wavelet function to expand the EM field components in orthonormal bases. However, this model shows similar characteristics as the FDTD model. Fujii and Hoefer [59] expanded the MRTD model by employing Daubechies' wavelet with only two vanishing moments [60] to three and four vanishing moments for time domain EM field modeling. Later on, other researchers used different scaling and wavelet functions such as Cohen-Daubechies-Feauveau (CDF) biorthogonal [61] and Coifman [62,63] scaling functions to present dispersion analysis, as a solution to scattering problems, etc.

Recently, Tong et al. [64] developed the MRTD model for the two-conductor lossless and lossy TL equations based on scaling functions of Daubechies' wavelet. They performed numerical dispersion and stability analysis of this model which made evident that the developed model shows better dispersion characteristics than the FDTD model. Moreover, considering the high vanishing moment of the scaling function

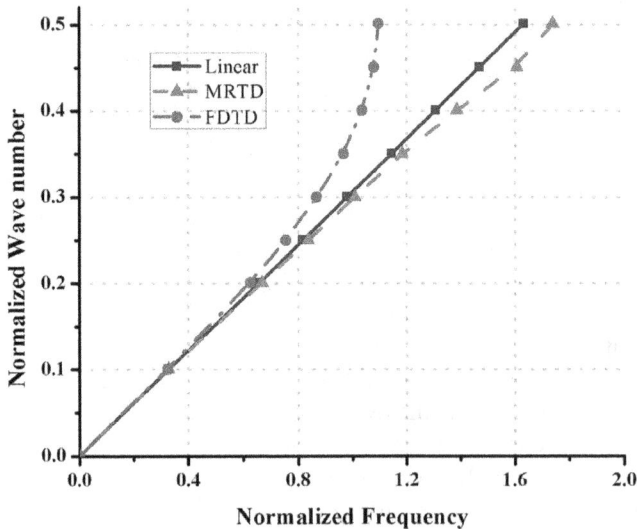

FIGURE 10.5 Dispersion curves of the MRTD model employing Battle-Lemarie scaling function and FDTD scheme with respect to ideal linear case adopted from [56].

provides more accurate results. Later they extended their work to multiconductor TLs terminated with linear loads [65].

10.2 ANALYSIS OF GRAPHENE INTERCONNECTS USING MRTD

10.2.1 ESC Model for MWCNT/MLGNR Interconnects

This section examines a MWCNT/MLGNR interconnect line corresponding to RLC model. Figure 10.6 depicts the ESC model of CMOS-driven coupled MWCNT/MLGNR interconnects. Parasitic capacitances of CMOS are C_{gd} and C_{diff}. The overall imperfect contact resistance, denoted by (R_{mc}), and the quantum resistance, denoted by (R_q), combine to form the average equivalent resistance, denoted by R_{lp}. The carrier scattering resistance R_{Sc_i} is expressed in terms of its value per unit length (p.u.l.). L_{M_x} is the magnetic inductance p.u.l., and L_{K_i} is the distributive line inductance. The shell coupling capacitance and the quantum capacitance are used in the calculation of the distributive line capacitance, which is written as (C_{Q_i}). The p.u.l. electrostatic capacitance is denoted by the symbol C_{E_i}. The parameters of distributed line with subscript i represent Line 1,..., Line $(N-1)$,Line N at $i = 1,2,3...(N-1),N$. The values of L_{K_i} and C_{Q_i} can be evaluated using recursive expressions [67,68], while the values of C_{E_i}, L_{M_i}, and the mutual inductance and coupling capacitance between coupled MWCNT/MLGNR lines can be obtained using standard tools like the Ansoft Maxwell field solvers [69]. Here the load capacitance is denoted by C_L.

It can be seen clearly that the ESC model for both MWCNT as well as MLGNR is same. So, the formulation of MRTD model for both MWCNT and MLGNR is similar. However, the parasitics of MWCNT and MLGNR are based on their physical properties that will show individual result analysis.

10.2.2 MRTD Model Formulation for MWCNT/MLGNR Interconnects

For N-coupled MWCNT/MLGNR interconnect lines, the MRTD model is formulated by employing Daubechies' scaling function with four vanishing moments $(D^{(4)})$.

10.2.2.1 Modeling of Mutually Coupled MWCNT/MLGNR Interconnects

The coupled VLSI MWCNT/MLGNR interconnects are expressed using the telegrapher's equations for the transverse EM mode [39]:

$$\frac{\partial V(x,t)}{\partial x} + RI(x,t) + L\frac{\partial I(x,t)}{\partial t} = 0 \tag{10.4}$$

$$\frac{\partial I(x,t)}{\partial x} + C\frac{\partial V(x,t)}{\partial t} = 0 \tag{10.5}$$

where **V** is voltage and **I** is current (equations (10.4) and (10.5)), which are expressed in $N \times 1$ column vector form $\begin{bmatrix} V_1 & V_2 & \cdots & V_N \end{bmatrix}^T$ and $\begin{bmatrix} I_1 & I_2 & \cdots & I_N \end{bmatrix}^T$. The parasitics of interconnect are represented in $N \times N$ matrix p.u.l., as given in equation (10.6).

(a)

(b)

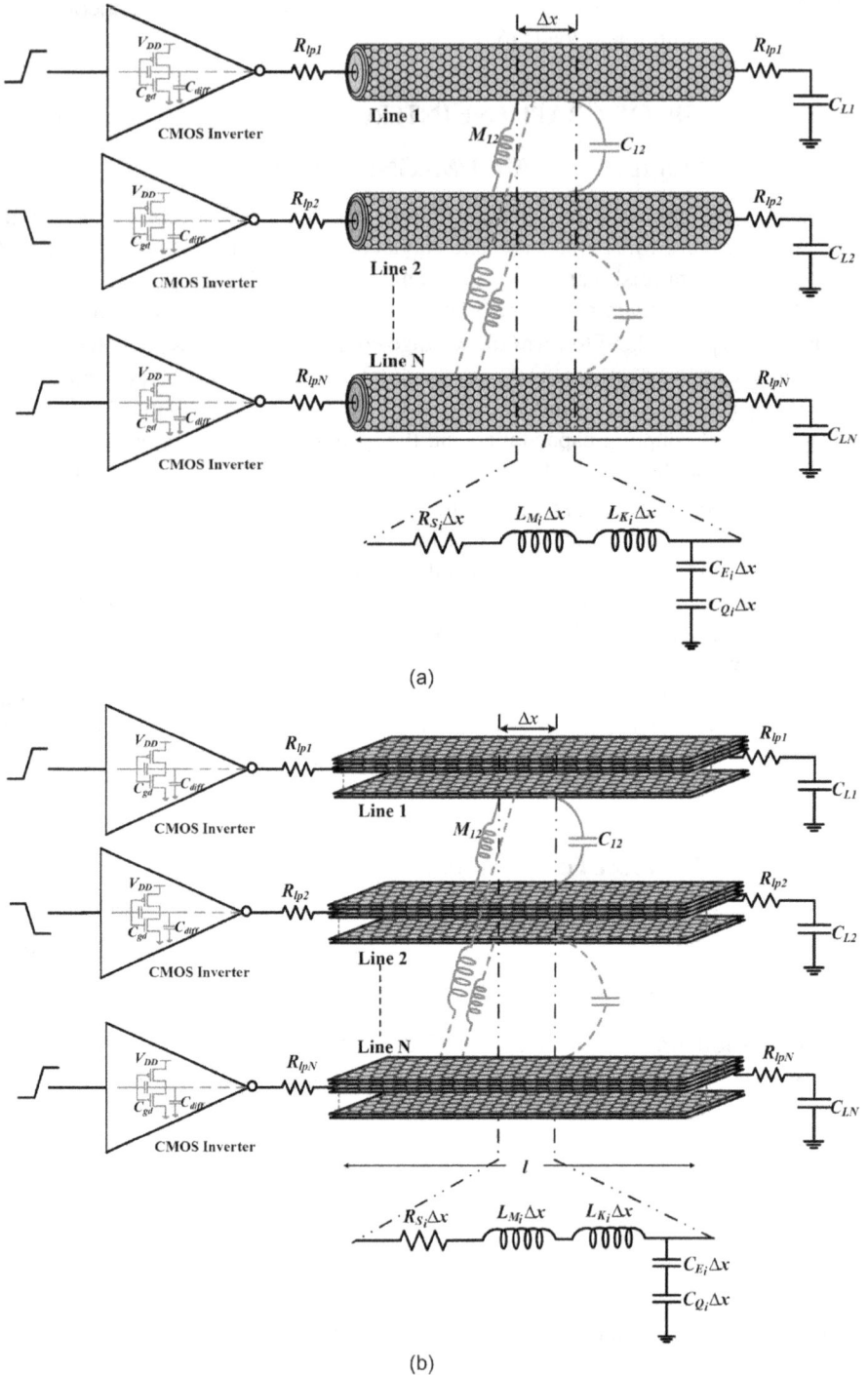

FIGURE 10.6 Mutually N-coupled (a) MWCNT [66] and (b) MLGNR interconnects driven by a nonlinear driver (CMOS), depicted as in ESC model's schematic.

$$R = diag\left[\; R_{S_1}\;\; R_{S_2}\;\; R_{S_3}, \cdot\; \cdot\; R_{S_{N-1}}\;\; R_{S_N}\; \right],$$

$$L = \begin{bmatrix} L_{K_1}+L_{M_1} & M_{12} & M_{13} & \cdots & M_{1(N-1)} & M_{1N} \\ M_{21} & L_{K_2}+L_{M_2} & M_{23} & \cdots & M_{2(N-1)} & M_{2N} \\ \cdots & \cdots & \cdots & \cdots & \cdots & \cdots \\ M_{N1} & M_{N2} & M_{N3} & \cdots & M_{N(N-1)} & L_{K_N}+L_{M_N} \end{bmatrix}$$

$$C = \begin{bmatrix} \left(\dfrac{1}{C_{Q_1}}+\dfrac{1}{C_{E_1}}\right)^{-1}+\displaystyle\sum_{y=2} C_{1y} & -C_{12} & -C_{13} & \cdots & 0 & 0 \\ -C_{21} & \left(\dfrac{1}{C_{Q_2}}+\dfrac{1}{C_{E_2}}\right)^{-1}+\displaystyle\sum_{y=1,3} C_{2y} & -C_{23} & \cdots & 0 & 0 \\ \cdots & \cdots & \cdots & \cdots & \cdots & \cdots \\ 0 & 0 & 0 & \cdots & -C_{N(N-1)} & \left(\dfrac{1}{C_{Q_N}}+\dfrac{1}{C_{E_N}}\right)^{-1}+\displaystyle\sum_{y=(N-1)} C_{Ny} \end{bmatrix}$$

(10.6)

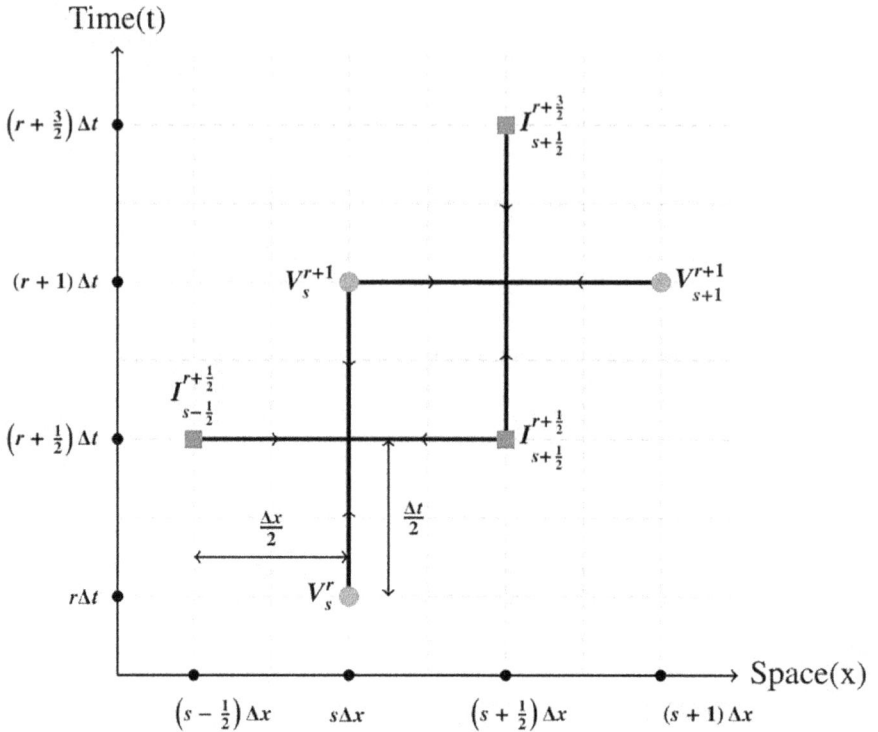

FIGURE 10.7   Relationship between the multiple segments of time (*t*) and space (*x*) adopted from [66].

The MRTD technique for solving telegrapher's equations achieves accuracy and stability by considering **V** and **I** separated by $\dfrac{\Delta x}{2}$ in space and $\dfrac{\Delta t}{2}$ in time, as shown in Figure 10.7.

A CMOS driver at $x = 0$ drives the *l*-long interconnect line, which ends at C_L at $x = l$. Figure 10.8 shows how the line is discretized into unknown coefficient of **I** and **V** nodes by splitting it uniformly into N_x segments of length $\Delta x = \dfrac{l}{N_x}$. Here the source current is I_0.

Equations (10.4) and (10.5) can be solved by expanding the **V** and **I** terms using the known functions ($\phi_s(x)$ and $h_r(t)$) and the unknown coefficients according to the method described in [54] as,

$$V(x,t) = \sum_{s,r=-\infty}^{+\infty} V_s^r \phi_s(x) h_r(t) \tag{10.7}$$

$$I(x,t) = \sum_{s,r=-\infty}^{+\infty} I_{s+\frac{1}{2}}^{r+\frac{1}{2}} \phi_{s+\frac{1}{2}}(x) h_{r+\frac{1}{2}}(t) \tag{10.8}$$

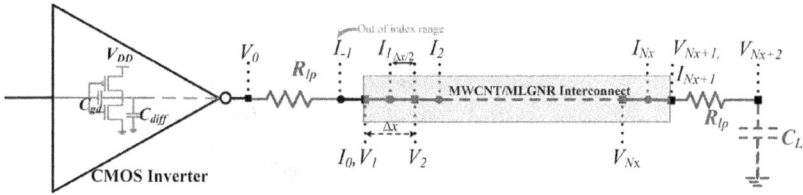

FIGURE 10.8 The MRTD model for a system of DIL is discretized spatially.

The coefficients for the scaling function expansion of V and I are denoted by V_s^r and $I_{s+\frac{1}{2}}^{r+\frac{1}{2}}$, respectively. Using the notation $t = r\Delta t$ and $x = s\Delta x$, we may express the relationship between the indices r and s in terms of time and space coordinates, respectively. The expression for $h_r(t)$ is: $h_r(t) = h\left(\dfrac{t}{\Delta t} - r\right)$

where $h(t)$ represents a pulse function that is expressed as

$$h(t) = \begin{cases} 1 \text{ for } |t| < \dfrac{1}{2} \\[2mm] \dfrac{1}{2} \text{ for } |t| = \dfrac{1}{2} \\[2mm] 0 \text{ for } |t| > \dfrac{1}{2} \end{cases} \tag{10.9}$$

The expression for $\phi_s(x)$ is given by

$$\phi_s(x) = \phi\left(\frac{x}{\Delta x} - s\right) \tag{10.10}$$

where $\phi(x)$ is known as Daubechies' scaling function.

To derive the MRTD method for equations (10.4) and (10.5), equations (10.11)–(10.14) are necessary [70]:

$$h_r(t), h_{r'}(t) = \delta_{r,r'}\Delta t \tag{10.11}$$

$$h_r(t), \frac{\partial h_{r'+\frac{1}{2}}(t)}{\partial t} = \delta_{r,r'} - \delta_{r,r'+1} \tag{10.12}$$

$$\phi_s(x), \phi_{s'}(x) = \delta_{s,s'}\Delta x \tag{10.13}$$

TABLE 10.1

Connection Coefficients $a(z)$

with Respect to $(D^{(4)})$ [59]

z	Connection Coeff. of $D^{(4)}$
1	1.3110340773
2	−0.1560100110
3	0.0419957460
4	−0.0086543236
5	0.0008308695
6	0.0000108999
7	0.0000000041

$$\phi_k(x), \frac{\partial \phi_{k'+\frac{1}{2}}(x)}{\partial x} = \sum_{z=-E_{ss}}^{E_{ss}-1} a(z)\delta_{s+z,s'} \tag{10.14}$$

In equation (10.13), for a basis function, E_{ss} stands for the effective support size, while $\delta_{r,r'}$, $\delta_{s,s'}$ stand for the Kronecker symbol. Connection coefficients, denoted by $a(z)$, are calculated by equation (10.15). Table 10.1 shows the values of $a(z)$ for $1 \leq z \leq E_{ss}$ when using $(D^{(4)})$ as the basis functions. For $z < 1$, $a(z)$ value can be obtained by the expression $a(-1-z) = -a(z)$, which is a symmetry relation, and for $z > E_{ss} \Rightarrow a(z) = 0$.

$$a(z) = \frac{1}{\pi} \int_0^\infty |\hat{\phi}(\lambda)|^2 \lambda \sin\lambda \left(z + \frac{1}{2}\right) d\lambda \tag{10.15}$$

Here the Fourier transform of a scaling function $\phi(x)$ is given by $\hat{\phi}(\lambda)$.

By using the Galerkin technique [54] with both the test functions $\phi_{s+\frac{1}{2}}(x)h_r(t)$ and $\phi_s(x)h_{r+\frac{1}{2}}(t)$ to equations (10.4) and (10.5), respectively, the iterative equations for the current and voltage may be obtained as follows:

$$I_{s+\frac{1}{2}}^{r+\frac{3}{2}} = Y_1 I_{s+\frac{1}{2}}^{r+\frac{1}{2}} - Y_2 \frac{\Delta t}{\Delta x} L^{-1} \sum_{z=1}^{E_{ss}} a(z)\left(V_{s+z}^r - V_{s-z+1}^r\right). \tag{10.16}$$

$$V_s^{r+1} = V_s^r - \frac{\Delta t}{\Delta x} C^{-1} \sum_{z=1}^{E_{ss}} a(z)\left(I_{s+z-\frac{1}{2}}^{r+\frac{1}{2}} - I_{s-z+\frac{1}{2}}^{r+\frac{1}{2}}\right). \tag{10.17}$$

where

$$Y_1 = \left(\frac{L}{R} + \frac{\Delta t}{2}\right)^{-1}\left(\frac{L}{R} - \frac{\Delta t}{2}\right) \text{ and } Y_2 = \left(1 + \frac{\Delta t}{2}RL^{-1}\right)^{-1}.$$

In equations (10.16) and (10.17), it is necessary to derive the terminal voltages V_1^{r+1} and V_{Nx+1}^{r+1} in addition to updating the **V** and **I** recursive equation near the terminals/boundaries. This is a critical part of the computation. Near the boundary, the **V** and **I** are represented as follows: V_z^{r+1}, V_{Nx+1-z}^{r+1} for $z = 2,3,\cdots,E_{ss}$ and $I_{z+\frac{1}{2}}^{r+\frac{1}{2}}$, $I_{Nx+1-z+\frac{1}{2}}^{r+\frac{1}{2}}$ for $z = 1,2,3,\cdots,E_{ss}-1$. There is a possibility that some of these **V** and **I** contain terms that are greater than the index range in (10.16) and (10.17).

Because of this, equations (10.16) and (10.17) need to be decomposed by applying the relation in [61] in order to update the iterative equation for voltages and currents in a way that satisfies the coefficients $a(z)$ given by.

$$\sum_{z=1}^{E_{ss}}(2z-1)a(z) = 1 \tag{10.18}$$

Substituting (10.18) into (10.17) results in

$$\sum_{z=1}^{E_{ss}}(2z-1)a(z)V_s^{r+1} = \sum_{z=1}^{E_{ss}}(2z-1)a(z)V_s^r$$

$$- \sum_{z=1}^{E_{ss}}\frac{\Delta t}{(2z-1)\Delta x}C^{-1}\left[(2z-1)a(z)\left(I_{s+z-\frac{1}{2}}^{r+\frac{1}{2}} - I_{s-z+\frac{1}{2}}^{r+\frac{1}{2}}\right)\right]. \tag{10.19}$$

It is possible to decompose equation (10.17) by looking at the equivalent terms with z, as shown here:

$$(2z-1)a(z)V_s^{r+1} = (2z-1)a(z)V_s^r$$

$$-(2z-1)a(z)\frac{\Delta t}{(2z-1)\Delta x}C^{-1}\left(I_{s+z-\frac{1}{2}}^{r+\frac{1}{2}} - I_{s-z+\frac{1}{2}}^{r+\frac{1}{2}}\right) \tag{10.20}$$

for $z = 1,2,3,\cdots,E_{ss}$.

In addition to this, improvements to equation (10.20) are possible by considering the relevant boundary conditions, as detailed in section.

10.2.2.2 Boundary Conditions are Incorporated Into the DIL System

Interconnect terminal conditions include the **V** and **I** equations at the driver and load. The voltage at the node (V_0) and the current at the node (I_0) are calculated using the nodal analysis.

$$I_0 = \frac{(V_0 - V_1)}{R_{lp}} \tag{10.21}$$

$$I_0 = C_{gd}\frac{dV_{in}}{dt} - \left(C_{gd} + C_{diff}\right)\frac{dV_0}{dt} + \left(I_h - I_e\right) \tag{10.22}$$

where $V_{in} = V_{GS}$ and $V_0 = V_{DS}$ and I_h and I_e represent current in the p-channel (pMOS) and n-channel (nMOS) metal oxide semiconductor transistor, respectively. The currents I_h and I_e are defined precisely by employing nth-power law model [46].

By applying the discretization to equation (10.21) and Galerkin method [54] to equation (10.22), we obtain

$$I_0^{r+1} = \frac{1}{R_{lp}}\left(V_0^{r+1} - V_1^{r+1}\right) \tag{10.23}$$

$$(\Delta x)(\Delta t)I_0^{r+1} = C_{gd}\left(\Delta x\right)\left(V_{in}^{r+1} - V_{in}^r\right) - \left(C_{gd} + C_{diff}\right)(\Delta x)\left(V_0^{r+1} - V_0^r\right)$$

$$+ (\Delta x)(\Delta t)I_h^{r+1} - (\Delta x)(\Delta t)I_e^{r+1} \tag{10.24}$$

$$\Downarrow$$

$$V_0^{r+1} = V_0^r + \left(\frac{C_{gd} + C_{diff}}{\Delta t}\right)^{-1}\left(\frac{C_{gd}}{\Delta t}\left(V_{in}^{r+1} - V_{in}^r\right) + I_h^{r+1} - I_e^{r+1} - I_0^r\right)$$

To calculate the voltage at the near-end terminal of this circuit, we substitute $s = 1$ in equation (10.17):

$$V_1^{r+1} = V_1^r - \frac{\Delta t}{\Delta x}C^{-1}\sum_{z=1}^{E_{ss}}a(z)\left(I_{z+\frac{1}{2}}^{r+\frac{1}{2}} - I_{-z+\frac{3}{2}}^{r+\frac{1}{2}}\right). \tag{10.25}$$

Equation (10.25) is decomposed by employing the steps from equation (10.20). Based on the decomposition, one can see that for the term $I_{-z+\frac{3}{2}}^{r+\frac{1}{2}}$ in equation (10.25) the subscript index range is exceeding for $z = 2, 3, 4, \cdots, E_{ss} - 1, E_{ss}$. In order to get over this constraint, a forward difference scheme is employed for the method of updating the voltage at the near-end terminal. As a result, the new terminal voltage, denoted by $\left(V_1^{r+1}\right)$, may be calculated as

$$V_1^{r+1} = V_1^r - \frac{\Delta t}{\Delta x}C^{-1}\sum_{z=1}^{L_s}2a(z)\left(I_{z+\frac{1}{2}}^{r+\frac{1}{2}} - I_0^{r+\frac{1}{2}}\right) \tag{10.26}$$

In equation (10.26), by substituting $I_0^{r+\frac{1}{2}} = \dfrac{I_0^r + I_0^{r+1}}{2}$ and I_0^{r+1}, V_0^{r+1} from (10.23), (10.24) we get

$$V_1^{r+1} = U_1 U_2 V_1^r + U_1 U_3 \left(\sum_{z=1}^{E_{ss}} a(z) \left(\frac{V_0^{r+1} + V_0^r}{R_{lp}} \right) - 2 \sum_{z=1}^{E_{ss}} a(z) I^{r+\frac{1}{2}}_{z+\frac{1}{2}} \right) \qquad (10.27)$$

where

$$U_1 = \left(1 + \frac{\Delta t}{\Delta x} C^{-1} R_{lp}^{-1} \sum_{z=1}^{E_{ss}} a(z) \right)^{-1}, \; U_2 = \left(1 - \frac{\Delta t}{\Delta x} C^{-1} R_{lp}^{-1} \sum_{z=1}^{E_{ss}} a(z) \right) \text{ and } U_3 = \frac{\Delta t}{\Delta x} C^{-1}.$$

Similarly, at the far-end terminal ($s = Nx + 1$) the load current (I_{Nx+1}) can be calculated by applying nodal analysis:

$$I_{Nx+1} = \frac{(V_{Nx+1} - V_{Nx+2})}{R_{lp}} \qquad (10.28)$$

$$I_{Nx+1} = C_L \frac{dV_{Nx+1}}{dt} \qquad (10.29)$$

By applying discretization to equation (10.28) and Galerkin method [54] to equation (10.20), we get

$$I_{Nx+1}^{r+1} = \frac{1}{R_{lp}} \left(V_{Nx+1}^{r+1} - V_{Nx+2}^{r+1} \right) \qquad (10.30)$$

$$V_{Nx+2}^{r+1} = V_{Nx+2}^r + \frac{\Delta t}{C_L} I_{Nx+1}^{r+1} \qquad (10.31)$$

Therefore, the updated voltage recursive equation at the far-end boundary is given by

$$V_{Nx}^{r+1} = U_1 U_2 V_{Nx}^r + U_1 U_3 \sum_{z=1}^{E_{ss}} a(z) \left(\frac{V_{Nx+2}^{r+1} + V_{Nx+2}^r}{R_{lp}} \right) + 2 U_1 U_3 \sum_{z=1}^{E_{ss}} a(z) I^{r+\frac{1}{2}}_{Nx+1-z+\frac{1}{2}} \qquad (10.32)$$

10.2.2.3   Voltage and Current Expressions at the Interior Points

It is necessary to eliminate terms whose indices are exceeding the index range for all nodes in between the terminals in order to derive and modify the recursive equations. This range applies to all nodes.

The decomposition of equation (10.17) can be done by considering the steps from equations (10.19) and (10.20) considering V_s^{r+1} as an example at $s = 2, 3, 4, \cdots, E_{ss} - 1, E_{ss}$.

$$a(1)V_s^{r+1} = a(1)V_s^r - a(1)\frac{\Delta t}{\Delta x}C^{-1}\left(I_{s+\frac{1}{2}}^{r+\frac{1}{2}} - I_{s-\frac{1}{2}}^{r+\frac{1}{2}}\right)$$

$$3a(2)V_s^{r+1} = 3a(2)V_s^r - 3a(2)\frac{\Delta t}{3\Delta x}C^{-1}\left(I_{s+\frac{3}{2}}^{r+\frac{1}{2}} - I_{s-\frac{3}{2}}^{r+\frac{1}{2}}\right)$$

$$\vdots$$

$$(2s-1)a(s)V_s^{r+1} = (2s-1)a(s)V_s^r - (2s-1)a(s)\frac{\Delta t}{(2s-1)\Delta x}C^{-1}\left(I_{2s-\frac{1}{2}}^{r+\frac{1}{2}} - I_{\frac{1}{2}}^{r+\frac{1}{2}}\right)$$

$$(2s+1)a(s+1)V_s^{r+1} = (2s+1)a(s+1)V_s^r$$

$$- (2s+1)a(s+1)\frac{\Delta t}{(2s+1)\Delta x}C^{-1}\left(I_{2s+\frac{1}{2}}^{r+\frac{1}{2}} - I_{-\frac{1}{2}}^{r+\frac{1}{2}}\right)$$

$$\vdots$$

$$(2E_{ss}-1)a(E_{ss})V_s^{r+1} = (2E_{ss}-1)a(E_{ss})V_s^r$$

$$- (2E_{ss}-1)a(E_{ss})\frac{\Delta t}{(2E_{ss}-1)\Delta x}C^{-1}\left(I_{s+E_{ss}-\frac{1}{2}}^{r+\frac{1}{2}} - I_{s-E_{ss}+\frac{1}{2}}^{r+\frac{1}{2}}\right)$$

$$(10.33)$$

The first s terms' indices are limited within the range shown by equation (10.33). In the MRTD approach, the derivation of iterative equations becomes impracticable when dealing with remaining $E_{ss} - s$ terms that have indices that are out of the bounds. We overcome this issue by trimming out the solutions to the equations whose indices are exceeding.

By combining the first s terms in equation (10.33), we obtain the updated iterative equation,

$$V_s^{r+1} = V_s^r - \left(\sum_{z=1}^s (2z-1)a(z)\right)^{-1} W_2\left(\sum_{z=1}^s a(z)\left(I_{s+z-\frac{1}{2}}^{r+\frac{1}{2}} - I_{s-z+\frac{1}{2}}^{r+\frac{1}{2}}\right)\right) \quad (10.34)$$

at $s = 2,3,\cdots,E_{ss}$

Both the voltages at interior nodes and near the load are represented by the modified iterative equation, which is illustrated in equations (10.36) and (10.37), respectively.

$$V_s^{r+1} = V_s^r - W_2\left(\sum_{z=1}^{E_{ss}} a(z)\left(I_{s+z-\frac{1}{2}}^{r+\frac{1}{2}} - I_{s-z+\frac{1}{2}}^{r+\frac{1}{2}}\right)\right). \quad (10.35)$$

at $s = E_{ss}+1, E_{ss}+2,\cdots,Nx-E_{ss}, Nx-E_{ss}+1.$

$$V_s^{r+1} = V_s^r - \left(\sum_{z=1}^{Nx-z+1} (2z-1)a(z) \right)^{-1} W_2 \left(\sum_{z=1}^{Nx-z+1} a(z) \left(I_{s+z-\frac{1}{2}}^{r+\frac{1}{2}} - I_{s-z+\frac{1}{2}}^{r+\frac{1}{2}} \right) \right) \quad (10.36)$$

at $z = Nx - E_{ss} + 2, Nx - E_{ss} + 3, \cdots, Nx$.

It is possible to update the current equations in the same way as the voltage equations are modified, with a few slight adjustments. All of the nodes of the current appear inside the terminals, as shown in Figure 10.8, indicating that these nodes are half-integer coordinates. For this reason, decomposing equation (10.16) in the same way as voltage iterative equations only updates the near-end terminal currents. Hence, the simplified recursive equation of currents near to the source are

$$I_{1+\frac{1}{2}}^{r+\frac{3}{2}} = Y_1 I_{1+\frac{1}{2}}^{r+\frac{1}{2}} - Y_2 \frac{\Delta t}{\Delta x} L^{-1} \left(\sum_{z=1}^{E_{ss}} a(z) \left(V_{z+1}^{r+1} - V_1^{r+1} \right) \right). \quad (10.37)$$

at $s = 1$ and,

$$I_{s+\frac{1}{2}}^{r+\frac{3}{2}} = Y_1 I_{s+\frac{1}{2}}^{r+\frac{1}{2}} - Y_2 \left(\sum_{z=1}^{s} (2z-1)a(z) \right)^{-1} \frac{\Delta t}{\Delta x} L^{-1} \left(\sum_{z=1}^{s} a(z) \left(V_{s+z}^{r+1} - V_{s-z+1}^{r+1} \right) \right). \quad (10.38)$$

at $s = 2, 3, \cdots, E_{ss}$.

Both the currents at interior points and near the load are represented by the modified iterative equation, which is illustrated in equations (10.39) and (10.40), respectively

$$I_{s+\frac{1}{2}}^{r+\frac{3}{2}} = Y_1 I_{s+\frac{1}{2}}^{r+\frac{1}{2}} - Y_2 \frac{\Delta t}{\Delta x} L^{-1} \left(\sum_{z=1}^{E_{ss}} a(z) \left(V_{s+z}^{r+1} - V_{s-z+1}^{r+1} \right) \right). \quad (10.39)$$

at $s = E_{ss} + 1, E_{ss} + 2, \cdots, Nx - E_{ss}, Nx - E_{ss} + 1$.

$$I_{s+\frac{1}{2}}^{r+\frac{3}{2}} = Y_1 I_{s+\frac{1}{2}}^{r+\frac{1}{2}} - Y_2 \left(\sum_{z=1}^{Nx-z+1} (2z-1)a(z) \right)^{-1} \frac{\Delta t}{\Delta x} L^{-1} \left(\sum_{z=1}^{Nx-z+1} a(z) \left(V_{s+z}^{r+1} - V_{s-z+1}^{r+1} \right) \right). \quad (10.40)$$

at $z = Nx - E_{ss} + 2, Nx - E_{ss} + 3, \cdots, Nx$.

A bootstrapping method is utilized in order to perform the analysis of the new updated iterative equations for $\mathbf{V}$ and $\mathbf{I}$. To begin, the equations of voltage (10.27), (10.32), and (10.38)–(10.40) are solved at a certain time with regard to the past p values of $\mathbf{V}$ and $\mathbf{I}$. Thereafter, the current equations (10.41)–(10.44) are solved with regard to the past values of $\mathbf{I}$ and $\mathbf{V}$. The output of the recursive equations developed by MRTD is stabilized by taking into consideration the Courant-Friedrichs-Lewy (CFL) stability criterion [61,64] as the governing condition.

$$\Delta t \leq \frac{q\Delta x}{\vartheta} \qquad (10.41)$$

Here, q is a courant number calculated using $q = \dfrac{1}{\sum_{z=1}^{E_{ss}} \Delta|a(z)|} = \dfrac{\vartheta \Delta t}{\Delta x}$, and the phase velocity of propagation along the line denoted by ϑ.

For a CFL to be stable, the propagation time over each cell must be greater than the time step.

10.3 EVALUATION OF THE DEVELOPED MRTD METHOD BY COMPARISON

The MRTD method that was proposed is evaluated by comparing the findings with the SPICE simulator, which is the standard for the industry. When driving the interconnect load, the symmetric CMOS driver must be taken into consideration. An interconnect line with a length of 1 mm is selected together with a realistic global interconnect structure for the 32 nm technology node. The DIL system's design parameters and interconnect dimensions are considered from [48].

It is assumed that R_{mc} per shell is 3.2 kΩ [48]. Employing a framework of feasible global interconnect design, we determine that the aspect ratio of line (T/W) should be 3, suggesting that each interconnect line is composed of three MWCNTs along its thickness. The line load capacitance is assumed to be 2 fF, and the inter-layer dielectric constant is set at 2.25.

10.3.1 DETAILED CROSSTALK ESTIMATION OF COUPLED MWCNT INTERCONNECTS

In this section, aggressor line is represented with line 1 and the victim line is represented with line 2 in Figure 10.9 when analyzing two mutually connected MWCNT interconnects. FDTD, SPICE, and the presented MRTD model study functional and dynamic crosstalk effects. Line 1 is switched from 0.9 V (V_{DD}) to 0 V while line 2 remains quiescent to study the functional crosstalk impact. Dynamic crosstalk is studied by switching both lines simultaneously. The far-end terminal on line 2 compares transient waveforms from the above circumstances (victim). Figure 10.9a–c presents line 2 (victim)'s function crosstalk, dynamic in-phase and out-of-phase transient response. As shown in Figures 10.9b and c, the conventional FDTD method's significant dispersion problems can cause peak undershoot/overshoot in line 2's response. The suggested MRTD model outperforms the FDTD model in accuracy because of its numerical dispersion advantages [58,59]. In Figure 10.9c, Miller coupling capacitance (C_{12}) causes signal transition to take longer during out-phase switching than in-phase switching. The new MRTD model matches SPICE and outperforms the standard FDTD approach for all input switching scenarios.

Table 10.2 shows the SPICE computational error for estimating the effect of dynamic crosstalk over line 2 (victim) for the proposed MRTD method. Testing the

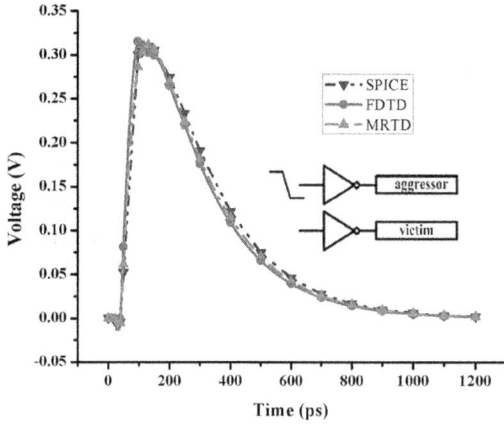

(a)

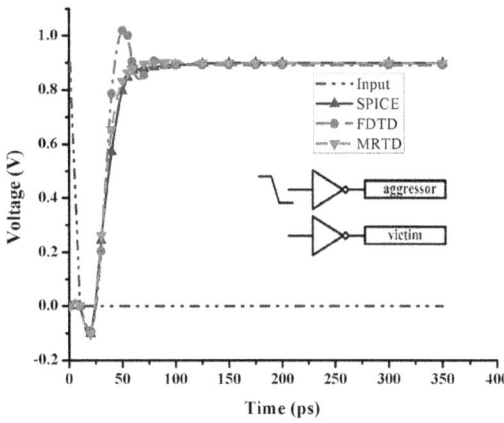

(b)

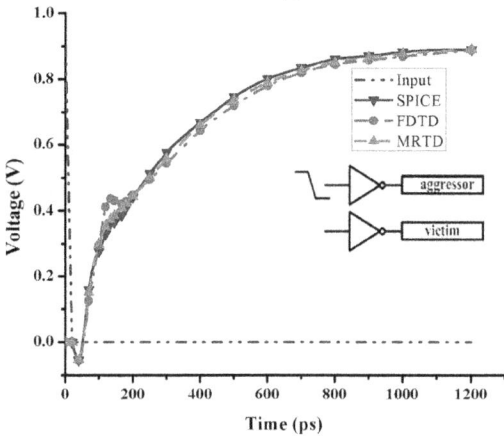

(c)

FIGURE 10.9 The transients of line 2 are compared in three different switching scenarios: (a) functional, (b) dynamic in-phase, and (c) dynamic out-of-phase switching, adopted from [66].

TABLE 10.2

Percentage Error in Measuring Dynamic Crosstalk Delay on Line 2 for Different Interconnect Lengths

	Dynamic Crosstalk Noise					
	In-Phase Delay (ps)			Out-of-Phase Delay (ps)		
Length of Interconnect (μm)	MRTD	SPICE	% of Error	MRTD	SPICE	% of Error
100	5.71	5.67	0.701	6.74	6.76	−0.297
200	6.366	6.31	0.879	10.92	10.81	1.007
300	7.05	6.99	0.851	15.02	14.86	1.065
400	8.95	8.87	0.894	20.736	20.54	0.945
500	10.974	10.87	0.948	26.71	26.49	0.824
600	13.07	12.9	1.301	35.05	34.59	1.312
700	15.11	14.93	1.191	45.43	44.86	1.255
800	17.2	16.95	1.453	56.1	55.41	1.229
900	19.15	18.87	1.462	66.57	65.67	1.35
1000	21.3	21.01	1.362	78.15	77.03	1.433

model for varied interconnect lengths yields an average inaccuracy of less than 2% for dynamic in-phase and out-of-phase delays.

10.3.2 CROSSTALK ESTIMATION OF COUPLED MLGNR INTERCONNECTS

The MRTD model is used to study the performance of coupled MLGNR interconnects by considering the smooth edges, that is, the mean free path (MFP) as a width-independent variable. Model validation is performed using the industry standard SPICE tool. Considering the width-independent MFPs, Figure 10.10 depicts the change in dynamic crosstalk delay with interconnect width for in-phase and out-of-phase transitions. Consider a rise/fall time of 10 ps for the input signal. The width of the interconnect can vary from 10 to 50 nm and the length of the interconnect is considered as 500 μm.

Finally, Table 10.3 displays the proposed model's corresponding computing effort in relation to SPICE. It has been noted that SPICE has a longer CPU run-time than the MRTD and FDTD schemes, but that MRTD is a little slower than FDTD due to MRTD's use of more iterations to achieve greater accuracy. As a result, accuracy and simulation time are trade-offs.

10.4   CONCLUSION

Because of their exceptional mechanical, electrical, and thermal qualities, MWCNT and MLGNR are being investigated for use in future applications of VLSI interconnect. This chapter offered an effective MRTD method for analyzing the inclusive crosstalk effects in a coupled MWCNT/MLGNR interconnect. It has been noticed that the MRTD approach is in close agreement with the result achieved using the

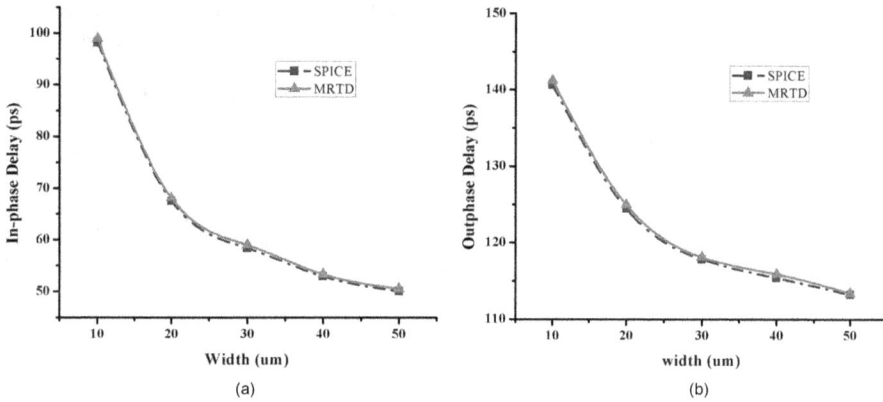

FIGURE 10.10 Estimation of dynamic crosstalk with the variation of width of MLGNR interconnect having smooth edges. (a) In-phase delay and (b) out-of-phase delay.

TABLE 10.3
Computational Performance Comparison
Adopted from [66]

Number of Coupled Lines	Computational Time (s)		
	MRTD	SPICE	FDTD
2	0.385	0.69	0.322
3	0.565	0.78	0.496

SPICE tool, and that it excels the FDTD in terms of accuracy. Additionally, it has been noted that the CPU run-time needed for the MRTD model is considerably smaller in comparison with SPICE.

REFERENCES

[1] ITRS, *ITRS Report on Interconnect Technology*, 2007. www.itrs2.net/itrs-reports.html
[2] C. Guoqing and E. G. Friedman, "An RLC interconnect model based on Fourier analysis," *IEEE Transactions on Computer-Aided Design of Integrated Circuits and Systems*, vol. 24, no. 2, pp. 170–183, 2005.
[3] F. Moll and M. Roca, *Interconnection Noise in VLSI Circuits*. Springer, New York, 2004.
[4] J. Nurmi, J. Isoaho, A. Jantsch, and H. Tenhunen, *Interconnect-Centric Design for Advanced SoC and NoC*. Springer, New York, 2005.
[5] H. B. Bakoglu, *Circuits, Interconnections, and Packaging for VLSI*. Addison Wesley, Boston, MA, 1990.
[6] C. R. Paul, *Analysis of Multiconductor Transmission Lines*. John Wiley & Sons, New York, 1994.
[7] R. Achar and M. S. Nakhla, "Simulation of high-speed interconnects," *Proceedings of the IEEE*, vol. 89, no. 5, pp. 693–728, 2001.

[8] J. H. Chern, J. Huang, L. Arledge, P. C. Li, and P. Yang, "Multilevel metal capacitance models for CAD design synthesis systems," *IEEE Electron Device Letters*, vol. 13, no. 1, pp. 32–34, 1992.

[9] T. Sakurai, "Approximation of wiring delay in MOSFET LSI," *IEEE Journal of SolidState Circuits*, vol. 18, no. 4, pp. 418–426, 1983.

[10] W. C. Elmore, "The transient response of damped linear networks with particular regard to wideband amplifiers," *Journal of Applied Physics*, vol. 19, no. 1, pp. 55–63, 1948.

[11] T. Sakurai, "Closed-form expressions for interconnection delay, coupling, and crosstalk in VLSIs," *IEEE Transactions on Electron Devices*, vol. 40, no. 1, pp. 118–124, 1993.

[12] Y. I. Ismail and E. G. Friedman, "Effects of inductance on the propagation delay and repeater insertion in VLSI circuits," *IEEE Transactions on Very Large Scale Integration (VLSI) Systems*, vol. 8, no. 2, pp. 195–206, 2000.

[13] D. Sylvester and K. Shephard, "Electrical integrity design and verification for digital and mixed-signal systems on a chip," In: *Tutorial International Conference on Computer Aided Design,* San Jose, CA, 2001.

[14] A. Vittal, L. H. Chen, M. Marek-Sadowska, K. P. Wang, and S. Yang, "Crosstalk in VLSI interconnections," *IEEE Transactions on Computer-Aided Design of Integrated Circuits and Systems*, vol. 18, no. 12, pp. 1817–1824, 1999.

[15] M. Sahoo, P. Ghosal, and H. Rahaman, "Modeling and analysis of crosstalk induced effects in multiwalled carbon nanotube bundle interconnects: An ABCD parameterbased approach," *IEEE Transactions on Nanotechnology*, vol. 14, no. 2, pp. 259–274, 2015.

[16] V. R. Kumar, B. K. Kaushik, and M. K. Majumder, "Graphene based on-chip interconnects and TSVs: Prospects and challenges," *IEEE Nanotechnology Magazine*, vol. 8, no. 4, pp. 14–20, 2014.

[17] Z. Junmou and E. G. Friedman, "Crosstalk modeling for coupled RLC interconnects with application to shield insertion," *IEEE Transactions on Very Large Scale Integration (VLSI) Systems*, vol. 14, no. 6, pp. 641–646, 2006.

[18] K. T. Tang and E. G. Friedman, "Interconnect coupling noise in CMOS VLSI circuits," In: *Proceedings of the 1999 International Symposium on Physical Design*. ACM, New York, 1999, pp. 48–53.

[19] L. Yin and L. He, "An efficient analytical model of coupled on-chip RLC interconnects," In: *Proceedings of the 2001 Asia and South Pacific Design Automation Conference*. ACM, New York, 2001, pp. 385–390.

[20] K. Agarwal, D. Sylvester, and D. Blaauw, "Modeling and analysis of crosstalk noise in coupled RLC interconnects," *IEEE Transactions on Computer-Aided Design of Integrated Circuits and Systems*, vol. 25, no. 5, pp. 892–901, 2006.

[21] B. K. Kaushik, S. Sarkar, R. P. Agarwal, and R. Joshi, "An analytical approach to dynamic crosstalk in coupled interconnects," *Microelectronics Journal*, vol. 41, no. 2, pp. 85–92, 2010.

[22] B. K. Kaushik and S. Sarkar, "Crosstalk analysis for a CMOS-gate-driven coupled interconnects," *IEEE Transactions on Computer-Aided Design of Integrated Circuits and Systems*, vol. 27, no. 6, pp. 1150–1154, 2008.

[23] E. Sicard and A. Rubio, "Analysis of crosstalk interference in CMOS integrated circuits," *IEEE Transactions on Electromagnetic Compatibility*, vol. 34, no. 2, pp. 124–129, 1992.

[24] A. Vittal and M. Marek-Sadowska, "Crosstalk reduction for VLSI," *IEEE Transactions on Computer-Aided Design of Integrated Circuits and Systems*, vol. 16, no. 3, pp. 290-298, 1997.

[25] A. B. Kahng and S. Muddu, "An analytical delay model for RLC interconnects," *IEEE Transactions on Computer-Aided Design of Integrated Circuits and Systems*, vol. 16, no. 12, pp. 1507–1514, 1997.

[26] Y. Cao, X. Huang, D. Sylvester, N. Chang, and C. Hu, "A new analytical delay and noise model for on-chip RLC interconnect," In: *International Electron Devices Meeting 2000: Technical Digest. IEDM* (Cat. No. 00CH37138). IEEE, New York, 2000, pp. 823–826.

[27] Y. I. Ismail, E. G. Friedman, and J. L. Neves, "Equivalent Elmore delay for RLC trees," *IEEE Transactions on Computer-Aided Design of Integrated Circuits and Systems*, vol. 19, no. 1, pp. 83–97, 2000.

[28] J. A. Davis and J. D. Meindl, "Compact distributed RLC interconnect models. I. Single line transient, time delay, and overshoot expressions," *IEEE Transactions on Electron Devices*, vol. 47, no. 11, pp. 2068–2077, 2000.

[29] J. A. Davis and J. D. Meindl, "Compact distributed RLC interconnect models-Part II: Coupled line transient expressions and peak crosstalk in multilevel networks," *IEEE Transactions on Electron Devices*, vol. 47, no. 11, pp. 2078–2087, 2000.

[30] M. Kuhlmann and S. S. Sapatnekar, "Exact and efficient crosstalk estimation," *IEEE Transactions on Computer-Aided Design of Integrated Circuits and Systems*, vol. 20, no. 7, pp. 858–866, 2001.

[31] K. Banerjee and A. Mehrotra, "Analysis of on-chip inductance effects for distributed RLC interconnects," *IEEE Transactions on Computer-Aided Design of Integrated Circuits and Systems*, vol. 21, no. 8, pp. 904–915, 2002.

[32] X. C. Li, J. F. Mao, and H. F. Huang, "Accurate analysis of interconnect trees with distributed RLC model and moment matching," *IEEE Transactions on Microwave Theory and Techniques*, vol. 52, no. 9, pp. 2199–2206, 2004.

[33] A. K. Palit, V. Meyer, W. Anheier, and J. Schloeffel, "ABCD modeling of crosstalk coupling noise to analyze the signal integrity losses on the victim interconnect in DSM chips," In: *18th International Conference on VLSI Design held jointly with 4th International Conference on Embedded Systems Design*. IEEE, New York, 2005, pp. 354–359.

[34] A. K. Palit, S. Hasan, and W. Anheier, "Decoupled victim model for the analysis of crosstalk noise between on-chip coupled interconnects," In: *2009 11th Electronics Packaging Technology Conference*. IEEE, New York, 2009, pp. 697–701.

[35] G. Zhou, L. Su, D. Jin, and L. Zeng, "A delay model for interconnect trees based on ABCD matrix," In: *Proceedings of the 2008 Asia and South Pacific Design Automation Conference*. IEEE Computer Society Press, New York, 2008, pp. 510–513.

[36] B. K. Kaushik and S. Sarkar, "Crosstalk analysis for a CMOS gate driven inductively and capacitively coupled interconnects," *Microelectronics Journal*, vol. 39, no. 12, pp. 1834–1842, 2008.

[37] T. Sakurai and A. R. Newton, "Alpha-power law MOSFET model and its applications to CMOS inverter delay and other formulas," *IEEE Journal of Solid-State Circuits*, vol. 25, no. 2, pp. 584–594, 1990.

[38] K. Yee, "Numerical solution of initial boundary value problems involving Maxwell's equations in isotropic media," *IEEE Transactions on Antennas and Propagation*, vol. 14, no. 3, pp. 302–307, 1966.

[39] C. R. Paul, "Incorporation of terminal constraints in the FDTD analysis of transmission lines," *IEEE Transactions on Electromagnetic Compatibility*, vol. 36, no. 2, pp. 85–91, 1994.

[40] J. A. Roden, C. R. Paul, W. T. Smith, and S. D. Gedney, "Finite-difference time-domain analysis of lossy transmission lines," *IEEE Transactions on Electromagnetic Compatibility*, vol. 38, no. 1, pp. 15–24, 1996.

[41] X. Li, J. Mao, and M. Swaminathan, "Accurate analysis of CMOS inverter driving transmission line based on FDTD," In: *2009 IEEE MTT-S International Microwave Symposium Digest*. IEEE, New York, 2009, pp. 1573–1576.

[42] D. K. Sharma, S. Mittal, B. Kaushik, R. Sharma, K. Yadav, and M. Majumder, "Dynamic crosstalk analysis in RLC modeled interconnects using FDTD method," In: *2012 Third International Conference on Computer and Communication Technology*. IEEE, New York, 2012, pp. 326–330.

[43] D. K. Sharma, B. K. Kaushik, and R. Sharma, "Signal integrity and propagation delay analysis using FDTD technique for VLSI interconnects," *Journal of Computational Electronics*, vol. 13, no. 1, pp. 300–306, 2014.

[44] V. R. Kumar, B. K. Kaushik, and A. Patnaik, "An accurate model for dynamic crosstalk analysis of CMOS gate driven on-chip interconnects using FDTD method," *Microelectronics Journal*, vol. 45, no. 4, pp. 441–448, 2014.

[45] V. R. Kumar, B. K. Kaushik, and A. Patnaik, "An accurate FDTD model for crosstalk analysis of CMOS-gate-driven coupled RLC interconnects," *IEEE Transactions on Electromagnetic Compatibility*, vol. 56, no. 5, pp. 1185–1193, 2014.

[46] T. Sakurai and A. R. Newton, "A simple MOSFET model for circuit analysis," *IEEE Transactions on Electron Devices*, vol. 38, no. 4, pp. 887–894, 1991.

[47] V. R. Kumar, B. K. Kaushik, and A. Patnaik, "Crosstalk noise modeling of multiwall carbon nanotube (MWCNT) interconnects using finite-difference time-domain (FDTD) technique," *Microelectronics Reliability*, vol. 55, no. 1, pp. 155–163, 2015.

[48] V. R. Kumar, B. K. Kaushik, and A. Patnaik, "Improved crosstalk noise modeling of MWCNT interconnects using FDTD technique," *Microelectronics Journal*, vol. 46, no. 12, pp. 1263–1268, 2015.

[49] V. R. Kumar, B. K. Kaushik, and A. Patnaik, "Crosstalk modeling with width dependent MFP in MLGNR interconnects using FDTD technique," In: *2015 IEEE International Conference on Electron Devices and Solid-State Circuits (EDSSC)*. IEEE, New York, 2015, pp. 138–141.

[50] Y. Agrawal and R. Chandel, "Crosstalk analysis of current-mode signalling-coupled RLC interconnects using FDTD technique," *IETE Technical Review*, vol. 33, no. 2, pp. 148–159, 2016.

[51] Y. Agrawal, M. G. Kumar, and R. Chandel, "Comprehensive model for high-speed current-mode signaling in next generation MWCNT bundle interconnect using FDTD technique," *IEEE Transactions on Nanotechnology*, vol. 15, no. 4, pp. 590–598, 2016.

[52] Y. Agrawal, M. G. Kumar, and R. Chandel, "A novel unified model for copper and MLGNR Interconnects using voltage-and current-mode signaling schemes," *IEEE Transactions on Electromagnetic Compatibility*, vol. 59, no. 1, pp. 217–227, 2016.

[53] A. Kumar, V. R. Kumar, and B. K. Kaushik, "Transient analysis of crosstalk induced effects in mixed CNT bundle interconnects using FDTD technique," *IEEE Transactions on Electromagnetic Compatibility*, vol. 61, no. 5, pp. 1621–1629, 2018.

[54] M. Krumpholz and L. P. B. Katehi, "MRTD: New time-domain schemes based on multiresolution analysis," *IEEE Transactions on Microwave Theory and Techniques*, vol. 44, no. 4, pp. 555–571, 1996.

[55] E. M. Tentzeris, R. L. Robertson, J. F. Harvey, and L. P. B. Katehi, "Stability and dispersion analysis of Battle-Lemarie-based MRTD schemes," *IEEE Transactions on Microwave Theory and Techniques*, vol. 47, no. 7, pp. 1004–1013, 1999.

[56] I. Massy and M. M. Ney, "A Hybrid MRTD-FDTD Technique for Efficient Field Computation," In: *Computational Electromagnetics-Retrospective and Outlook*. Springer, New York, 2015, pp. 245–278.

[57] M. Fujii and W. J. R. Hoefer, "Multiresolution analysis similar to the FDTD method derivation and application," *IEEE Transactions on Microwave Theory and Techniques*, vol. 46, no. 12, pp. 2463–2475, 1998.

[58] S. Grivet-Talocia, "On the accuracy of Haar-based multiresolution time-domain schemes," *IEEE Microwave and Guided Wave Letters*, vol. 10, no. 10, pp. 397–399, 2000.

[59] M. Fujii and W. J. R. Hoefer, "Dispersion of time domain wavelet Galerkin method based on Daubechies' compactly supported scaling functions with three and four vanishing moments," *IEEE Microwave and Guided Wave Letters*, vol. 10, no. 4, pp. 125–127, 2000.

[60] W. C. Young, M. L. Yong, H. R. Keuk, G. K. Joon, and C. S. Chull, "Wavelet-Galerkin scheme of time-dependent inhomogeneous electromagnetic problems," *IEEE Microwave and Guided Wave Letters*, vol. 9, no. 8, pp. 297–299, 1999.

[61] T. Dogaru and L. Carin, "Multiresolution time-domain algorithm using CDF biorthogonal wavelets," *IEEE Transactions on Microwave Theory and Techniques*, vol. 49, no. 5, pp. 902–912, May 2001.

[62] W. Xingchang, L. Erping, and L. Changhong, "A new MRTD scheme based on Coifman scaling functions for the solution of scattering problems," *IEEE Microwave and Wireless Components Letters*, vol. 12, no. 10, pp. 392–394, 2002.

[63] A. Alighanbari and C. D. Sarris, "Dispersion properties and applications of the Coifman scaling function based S-MRTD," *IEEE Transactions on Antennas and Propagation*, vol. 54, no. 8, pp. 2316–2325, 2006.

[64] Z. Tong, L. Sun, Y. Li, and J. Luo, "Multiresolution time-domain scheme for terminal response of two-conductor transmission lines," *Mathematical Problems in Engineering*, vol. 2016, p. 15, 2016.

[65] Z. Tong, L. Sun, Y. Li, L. D. Angulo, S. G. Garcia, and J. Luo, "Multiresolution time domain analysis of multiconductor transmission lines terminated in linear loads," *Mathematical Problems in Engineering*, vol. 2017, p. 15, 2017.

[66] S. Rebelli and B. R. Nistala, "A multiresolution time domain (MRTD) method for crosstalk noise modeling of CMOS-gate-driven coupled MWCNT interconnects," *IEEE Transactions on Electromagnetic Compatibility*, vol. 62, no. 2, pp. 521–531, 2020.

[67] D. Das and H. Rahaman, *Carbon Nanotube and Graphene Nanoribbon Interconnects*. CRC Press, New York, 2014.

[68] M. S. Sarto and A. Tamburrano, "Single-conductor transmission-line model of multiwall carbon nanotubes," *IEEE Transactions on Nanotechnology*, vol. 9, no. 1, pp. 82–92, 2010.

[69] Maxwell 2D Student Version, "Ansoft corp." 2005.

[70] G. W. Pan, *Wavelets in Electromagnetics and Device Modeling*, vol. 159. John Wiley & Sons, New York, 2003.

Index

Note: **Bold** page numbers refer to tables and *italic* page numbers refer to figures.

For Product Safety Concerns and Information please contact our EU
representative GPSR@taylorandfrancis.com
Taylor & Francis Verlag GmbH, Kaufingerstraße 24, 80331 München, Germany

www.ingramcontent.com/pod-product-compliance
Lightning Source LLC
Chambersburg PA
CBHW060258220326
41598CB00027B/4151

* 9 7 8 1 0 3 2 3 6 3 8 2 0 *